The Cambridge Technical Series

General Editor: P. Abbott, B.A.

ALTERNATING CURRENTS IN THEORY AND PRACTICE

ALTERNATING CURRENTS

IN

THEORY AND PRACTICE

BY

W. H. N. JAMES, A.R.C.Sc. (Lond.), A.M.I.E.E.

Lecturer in Electrical Engineering, Bradford Technical College

Cambridge :

at the University Press

1916

CAMBRIDGE
UNIVERSITY PRESS

University Printing House, Cambridge CB2 8BS, United Kingdom

Cambridge University Press is part of the University of Cambridge.

It furthers the University's mission by disseminating knowledge in the pursuit of education, learning and research at the highest international levels of excellence.

www.cambridge.org
Information on this title: www.cambridge.org/9781316606964

© Cambridge University Press 1916

First published 1916
First paperback edition 2016

A catalogue record for this publication is available from the British Library

ISBN 978-1-316-60696-4 Paperback

PREFACE

THOUGH there are numerous books dealing with the Principles of Alternating Currents, the writer has on many occasions, when recommending a text book to students, found a difficulty in selecting one which, while giving a sufficiently full account of the laws governing the flow of alternating currents in circuits, also includes an account of the several types of alternating current machines.

It is hoped that the present book will fill the gap, since, while the fundamental principles of the alternating current circuit are adequately considered, sufficient matter concerning machines and appliances is given to enable readers who have worked through the book to proceed at once to a specialised study of the particular machine or appliance in which they are more immediately interested.

Many students, in fact the majority, desire to take up the study of alternating currents with a very inadequate knowledge of mathematics. In view of this fact, which is to be greatly regretted and which constitutes one of the greatest difficulties of students and teachers alike, the calculus has not been used in the body of the book; it has, however, been used in a few footnotes, mainly with the idea of showing students what excellent examples of the use of elementary calculus are furnished by the subject under consideration.

The writer would impress on students the great importance of obtaining a considerable and exact knowledge of the laws governing current flow in simple circuits; they will then find that the study of machines, in which these laws occur again and again, is very much facilitated.

In conclusion the writer wishes to tender his best thanks to the many firms who have kindly supplied information and diagrams concerning their products. He also desires to point out that the notation recommended by the International Electrotechnical Commission has been used throughout.

W. H. N. J.

8 *January* 1916

CONTENTS

CHAP.   PAGE

I. PRELIMINARY CONSIDERATIONS 1

II. INDUCTANCE 29

III. THE FLOW OF SINGLE PHASE ALTERNATING CURRENTS IN
CIRCUITS POSSESSING RESISTANCE, INDUCTANCE AND
CAPACITY 40

IV. POWER IN ALTERNATING CURRENT CIRCUITS . . . 62

V. MULTIPHASE CURRENTS 79

VI. INSTRUMENTS FOR USE ON ALTERNATING CURRENT CIRCUITS . 98

VII. ALTERNATORS 142

VIII. STATIC TRANSFORMERS 188

IX. INDUCTION MOTORS 224

X. CONVERTING PLANT 258

XI. SWITCHGEAR AND PROTECTIVE APPLIANCES; HIGH TENSION
TRANSMISSION 298

INDEX 351

Fig. 140 . . . . between pages 204–5

„ 167 . . . . „ „ 242–3

„ 219 . . . . „ „ 320–1

„ 220 . . . . „ „ 320–1

CHAPTER I

PRELIMINARY CONSIDERATIONS

When a conductor moves across a magnetic field in such a direction as to cut the magnetic lines representing that field, an electromotive force (E.M.F.) is produced within it, and, if the conductor forms part of a closed circuit, a current will, in general, result. Consider the E.M.F. produced in a conductor revolving at a uniform rate about an axis parallel to itself and at right angles

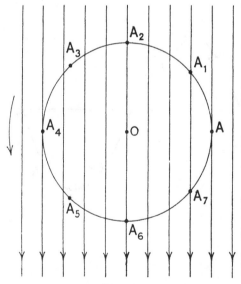

Fig. 1.

to the magnetic lines representing a uniform magnetic field. The E.M.F. in such a conductor is, when expressed in absolute units, numerically equal to the rate of cutting of magnetic lines and it is evident from an inspection of Fig. 1 that this rate will

vary considerably during one revolution of the conductor, being
a minimum for the position A and a maximum for the position A_2.

Consider the conductor starting from A, it is then moving
parallel to the magnetic lines, the rate of cutting (and consequently
the E.M.F.) being zero; as it reaches the position A_1 it is cutting
across the magnetic lines at an ever increasing rate giving a
greater and greater E.M.F., both these quantities attaining their
maximum values when the conductor reaches the position A_2
and is moving at right angles to the magnetic lines; then, as the
conductor passes on towards A_3, the pressure generated commences
to fall off, reaching a zero value when the conductor is at A_4 and
is again moving parallel to the magnetic lines.   During the passage
of the conductor from A_4 to A a similar set of changes to the above
is passed over but the pressures produced are in the opposite
direction.

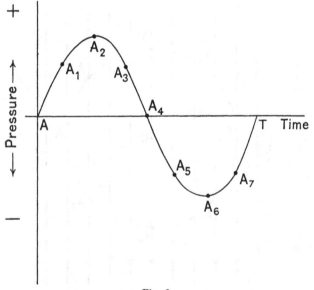

Fig. 2.

The complete set of changes which the E.M.F. in the conductor
undergoes will be best seen by reference to the curve in Fig. 2,
where time is plotted along the axis of abscissae, the distance
AT representing the time taken by the conductor in making one

revolution, and the pressures in the conductor are plotted as ordinates. The points marked A, A_1, A_2, etc. correspond to the points similarly marked in Fig. 1.

To investigate the true shape of the pressure curve obtained under the conditions dealt with above, we notice that the direction of motion of the conductor may at any place be resolved into two components, one parallel to the magnetic lines and the other at right angles to them, and it is the latter only which is effective in producing pressure.

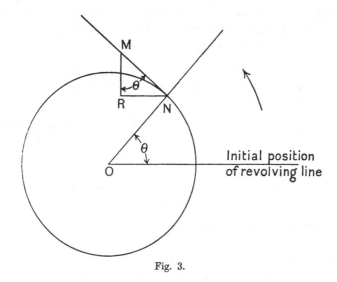

M

θ

R   N

θ

O

Initial position
of revolving line

Fig. 3.

In Fig. 3 if the line NM be drawn to some convenient scale to represent the total velocity of the point N, then to the same scale NR represents the velocity at right angles to the lines and MR the velocity parallel to them. Let the maximum value of the E.M.F. produced in the conductor be E_m, this is of course produced when the conductor is cutting across the lines with the maximum velocity represented by NM and will occur in the position A_2 (see Fig. 1) when the total velocity of the conductor and the velocity across the lines will be equal.

Since the E.M.F. will be proportional to the rate of cutting lines, its value at the point N will be $E_m \times \dfrac{NR}{NM} = E_m \times \sin\theta$,

θ being the angle which the line has moved through from the position of zero E.M.F.

If the conductor is making f revolutions per second and the time t has elapsed since the conductor passed the zero position, then $\theta = 2\pi ft$ or the instantaneous value of the pressure is

$$e = E_m \sin 2\pi ft = E_m \sin \omega t, \text{ where } \omega = 2\pi f.$$

Example. A loop of wire of the dimensions shown revolves about the axis AB (which is at right angles to the lines of the magnetic field) at 50 revolutions per second. If the strength of the field is 100 lines per square centimetre, calculate the maximum value of E.M.F. produced in the loop and draw a curve connecting t and e for one revolution.

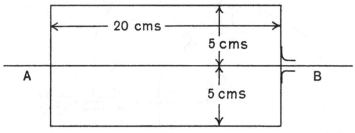

Fig. 4.

Peripheral velocity of loop $= \pi \times d \times n$

$$= 3.14 \times 10 \times 50$$
$$= 1570 \text{ cms. per second.}$$

Maximum pressure produced in both sides of loop

$$= 1570 \times 20 \times 2 \times 100 \text{ absolute units}$$
$$= \frac{1570 \times 20 \times 2 \times 100}{10^8} \text{ volt}$$
$$= .0628 \text{ volt.}$$

The pressure at any time t will be given by the equation

$$e = .0628 \sin 2\pi ft.$$

Thus when $t = .002$ second,

$$e = .0628 \sin (6.28 \times 50 \times .002) = .0628 \sin .628$$
$$\text{(angle is in circular measure)}$$
$$= .0628 \times .588 = .0369 \text{ volt.}$$

In a similar manner other points may be obtained and the curve plotted (see Fig. 5).

t (secs.)	0	·001	·002	·005	·008	·009
e	0	·0194	·0369	·0628	·0369	·0194
t (secs.)	·010	·011	·012	.015	·018	·019
e	0	−·0194	−·0369	−·0628	−·0369	−·0194

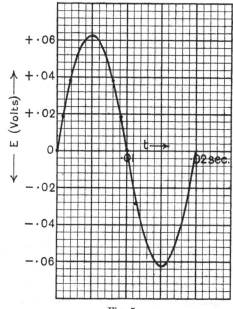

Fig. 5.

The student may not be impressed by the magnitude of the pressure in the above example, but he should remember that pressures of any desired magnitude may be produced by using a sufficient size and number of turns and a strong magnetic field (which can be obtained when iron cores are used).

Though usual it is not essential that the radian be used as the unit angle, the degree may be used if necessary and if so a slight modification in one of the constants results. Since 2π radians is equivalent to 360° we get

$$e = E_m \sin 360ft.$$

The student will often come across expressions such as

$$e = E_m \cos 2\pi ft,$$

but a little thought will show that this is obtained by measuring time from the instant of maximum value (in the positive direction) instead of from the instant of zero value (increasing positively).

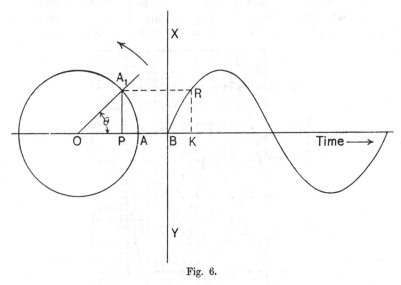

Fig. 6.

The curve shown in Fig. 5 is known as a sine or sinusoidal curve and may also be obtained by a graphical method. In Fig. 6 let the line OA (whose length to some convenient scale represents the maximum value of a current or pressure) rotate round O at the rate of f revolutions per second (corresponding to a frequency of f cycles per second). When A has moved to the point A_1 we have

$$PA_1 = A_1O \sin \theta = AO \sin 2\pi ft,$$

where t is the time occupied in moving through the angle θ, and this expression is also equal to the instantaneous value of the quantity whose maximum value is OA. If therefore we plot the lengths of the projections of OA on such a line as XY as ordinates, and the corresponding times which have elapsed since the revolving line occupied the position OA as abscissae, we shall have a sine wave similar to that obtained above. Thus in the diagram when the line has rotated through θ degrees the length of the projection is RK, and the corresponding time is BK.

Definitions.

A **cycle** or **period** indicates one complete set of changes both in magnitude and sign of the quantity concerned, commencing at a definite part of one wave and extending to the corresponding part of the next. Thus in Fig. 7 the portion of the curve shown from A to B constitutes a complete cycle, similarly the portion

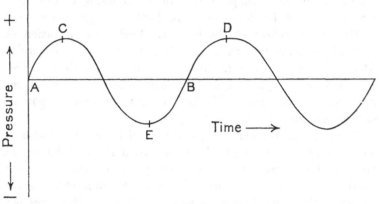

Fig. 7.

of the curve from C to D also forms a complete cycle. Occasionally the term alternation is met with and by this is meant half a cycle, usually the half extending from the position of positive maximum to the position of negative maximum or *vice versâ*. Thus from C to E or from E to D constitutes an alternation.

Frequency or **Periodicity** indicates the rate at which cycles are performed and is usually expressed in cycles per second though occasionally the number of alternations per minute is stated. A pressure having a frequency of 50 cycles per second has also a frequency of 6000 alternations per minute. In commercial forms of alternators an individual conductor has to pass, or be passed, by two adjacent poles of opposite polarity for a complete cycle to occur, and thus the number of poles passing in front of a given conductor per second, when divided by two, gives the frequency of the pressure (and current) generated in cycles per second.

While dealing with the definition of frequency a few words concerning the frequencies in common use may not be out of place.

For each particular application of alternating currents a certain range of frequencies gives the best results and if, as is usually the case, a certain source of supply is used for several purposes a compromise has to be arrived at.   An ordinary alternating current supply has commonly to feed incandescent lamps, arc lamps and induction motors; for incandescent lamps the frequency should not be too low otherwise the lamps, especially if for low candle power and high pressure, will flicker owing to the small heat capacity of the filament combined with the cyclical variation of the heating effect of the current.   Mechanical vibration of parts of the filament may also be looked for at low frequencies which will tend to reduce the life of the lamp.   Both these sources of trouble are small at 50 cycles per second and this, or any higher frequency (within reasonable limits), may be looked upon as suitable for incandescent lamps.

In arc lamps the flickering is worse and is very noticeable at 50 cycles per second (especially when bright moving objects are viewed), it is considerably less at 75 cycles per second but the minimum frequency for really satisfactory operation of arc lamps on alternating current circuits may be taken as 100 cycles per second.

Induction motors operate most successfully at from 40 to 60 cycles per second, and alternating current commutator motors with 15 to 25 cycles per second.

Summing up we may say that for supply systems dealing with a combined motor and lighting load 50 cycles per second is the most suitable frequency and this will probably become standard, while for systems dealing with lighting only 100 cycles per second would be better.   On the other hand when a station is installed for traction purposes, alternating current commutator motors being used, 25 cycles per second will prove most useful.   In the past many frequencies have been used, both in this country and abroad, but from practically every point of view there is a good deal to be said for the standardisation of a few frequencies such as those mentioned above.

Phase and **phase difference.**   Students of electrical engineering who have not had the opportunity of acquiring a fair knowledge of mathematics or physics will, in all probability, find the ideas of phase and phase difference quite new to them and yet, so far as many alternating current problems are concerned, phase difference is the all-important quantity, more important even than frequency or magnitude.

Information concerning the phase of a quantity may be given by stating the part of the cycle through which the quantity is passing at a specified time; thus we may say that a certain quantity is passing through its zero value, increasing positively, when $t = 0$.

In Fig. 8 two alternating pressure waves are shown, having equal frequencies and magnitudes, yet they are represented by curves having a considerable difference in position. The curve shown by the continuous line has the phase mentioned above, while the curve shown by the dotted line has a phase such that it is passing through its maximum positive value when $t = 0$, and

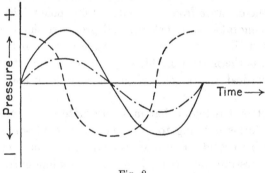

Fig. 8.

even at this early stage it is desirable to lay stress on the fact that these two waves, apparently so similar, might have very different effects when flowing in a circuit solely on account of their different phases. The curves shown by the continuous and chain lines, though differing in magnitude, have the same phase and are said to be "in phase" with each other.

The cases quoted above are instances where information concerning the absolute phase is given; as a matter of fact the difference of phase between two quantities is usually of far greater importance in practical problems, and from this point of view the pressures shown in Fig. 8 may be said to differ in phase by a quarter of a period, the dotted curve leading (since it passes through each part of the cycle before the other) and the continuous curve lagging. Instead of using the term "quarter of a period" we may say that the phase difference between the two waves is 90° or $\frac{\pi}{2}$ radians, since we look upon the complete cycle as consisting of 360° or 2π radians (think of the derivation of the wave from the rotating line as represented in Fig. 6).

Wave forms. In our preliminary consideration of the production of an alternating current, we studied the motion of a conductor rotating in a uniform magnetic field and in this case we obtained a curve whose magnitude varied as the sine of the angle through which the conductor had rotated. If we had a conductor moving at a uniform rate across a magnetic field whose distribution was sinusoidal (that is one whose strength at any point is proportional to the sine of the angle from the position of zero field strength, the distance from the centre of one pole to the centre of the next being taken as 180 electrical degrees) we should also have obtained a similar wave form. This wave form, it may be noted in passing, is the most desirable in practice, lending itself readily to mathematical treatment and giving simplicity and safety in operation.

In practice it is found that most alternators and transformers give wave forms differing to some extent, and in some cases to a considerable extent, from a sinusoidal wave, and even when a sinusoidal pressure wave is used the corresponding current waves may be considerably distorted as, for instance, when current is sent through a choking coil having an iron core.

If we had concentrated coils on an armature passing in front of poles of ordinary shape a flat-topped wave would result as in Fig. 9(a); again, if the turns were distributed somewhat, each turn in itself would still give a flat-topped wave but the pressures in the different turns would have slight phase differences and so would not exactly overlap and a wave form as shown in Fig. 9(b) would result.

Pressure waves approximating in shape to sine waves can be obtained in alternators by using windings which are to a certain extent distributed, in conjunction with suitably shaped pole pieces.

It will be realised then that in practice we may meet waves of varied, and at times complicated, shapes, they will, however, have certain properties in common amongst which may be mentioned:

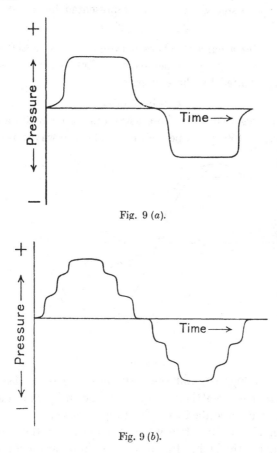

Fig. 9 (*a*).

Fig. 9 (*b*).

(1) The positive and negative halves of any wave are in general similar.

(2) The waves are always single valued, that is to say, for any value of the time there is but one corresponding value of the current or pressure as the case may be.

It is interesting and instructive to note that all alternating current wave forms, no matter how complicated they may be (providing always that they fulfil the two conditions mentioned above), can be represented, in accordance with Fourier's Theorem, by the superposition of a series of sine waves of various frequencies, magnitudes and phases.

Thus a pure sine wave may be represented by the expression

$$i = 200 \sin 314t,$$

which indicates a current whose maximum value is 200 amperes at a frequency of 50 cycles per second. Let us superpose on this a wave represented by the expression

$$i_1 = 30 \sin 942t,$$

which indicates a current whose maximum value is 30 amperes at a frequency of 150 cycles per second. The two waves are plotted

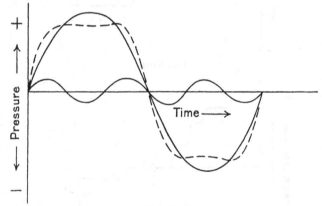

Fig. 10.

separately in Fig. 10, and the wave resulting from their super-position (that is from their addition) is shown by the dotted line (which may be described as a flat-topped curve).

In Fig. 11 the same two waves are added but the position of the one having the higher frequency (the third harmonic) has been altered relative to the other, the resultant wave this time being decidedly peaked.

Representation of Alternating Currents and Pressures.

In order to completely specify an alternating current (or pressure) information should be given on each of the following points:

(*a*) Wave form. This is usually taken to be sinusoidal and may always be assumed as such unless distinct mention is made to the contrary.

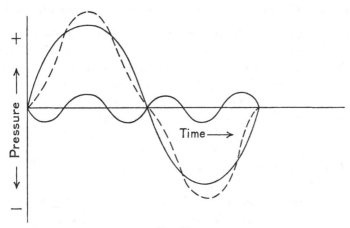

Fig. 11.

(*b*) Frequency. In considering a method of representing alternating quantities frequency is not perhaps of fundamental importance because, in the first place, the several currents and pressures to be dealt with in any problem will usually be of the same frequency, and, further, the behaviour of these currents and pressures will not be directly dependent on the frequency.

(*c*) Magnitude. This will always be of great importance in dealing with exact problems.

(*d*) Phase. A method of representation of alternating quantities is quite useless unless it is such as will give full information in regard to the relative phases of the several quantities concerned in a problem.

Three methods are in use which give information on the whole or the more important of the above points, they comprise:

(1) The use of trigonometrical formulae,

(2) The use of wave diagrams,

(3) The use of vector diagrams.

Trigonometrical formulae. Consider the formula

$$e = E_m \sin \omega t,$$

E_m and ω being suitable constants. This expression gives us information about wave form (*i.e.* that the pressure wave is sinusoidal), it tells us that the maximum value is E_m, that the frequency is $\dfrac{\omega}{2\pi}$ cycles per second and, finally, it gives us information about the phase of the pressure by indicating that the curve is passing through its zero value, increasing positively, when $t = 0$.

The modification of the above formula to indicate other pressures having different maximum values or frequencies is simple, but if it is desired to express pressures having different phases the matter is rather more complicated.

Consider the pressure expressed above to be indicated by E_1 in Fig. 12, then a pressure leading this by $\phi°$ would be produced in a conductor E_2 which is $\phi°$ in front of E_1 and which maintains this position as the two conductors go round and round. Evidently then the pressure in the conductor E_2 can be represented by the expression

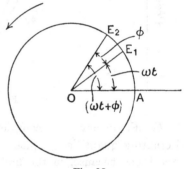

Fig. 12.

$$e_2 = E_{2m} \sin (\omega t + \phi).$$

Similarly a pressure lagging behind the first can be represented by

$$e_3 = E_{3m} \sin (\omega t - \phi).$$

Example. Express by trigonometrical formulae the following alternating currents and pressures.

(1) A pressure having a maximum value of 200 volts, a frequency of 50 cycles per second and which is passing through its zero value, increasing positively, when $t = 0$.

(2) A current having a maximum value of 100 amperes, a frequency of 50 cycles per second and which passes through its maximum positive value when $t = 0$.

(3) A pressure having a maximum value of 80 volts, a frequency of 50 cycles per second and lagging $\frac{1}{8}$ of a period behind (1).

(4) A current similar to (2) but having a frequency of 100 cycles per second.

(1) $e_1 = 200 \sin 2\pi \cdot 50 \cdot t = 200 \sin 314t.$

(2) $i_1 = 100 \sin \left(314t + \dfrac{\pi}{2}\right)$ or $100 \cos 314t.$

(3) $e_2 = 80 \sin \left(314t - \dfrac{\pi}{4}\right).$

(4) $i_2 = 100 \sin \left(628t + \dfrac{\pi}{2}\right)$ or $100 \cos 628t.$

The radian is the unit angle used in the above expressions but the degree may be used if desired, remembering that 360 degrees are equivalent to 2π radians, the answers then being

(1) $e_1 = 200 \sin (360 \cdot 50 \cdot t)° = 200 \sin 18000t°.$

(2) $i_1 = 100 \sin (360 \cdot 50 \cdot t + 90)° = 100 \sin (18000t + 90)°$
$ $ or $100 \cos 18000t°.$

(3) $e_2 = 80 \sin (18000t - 45)°.$

(4) $i_2 = 100 \sin (360 \cdot 100 \cdot t + 90)° = 100 \sin (36000t + 90)°$
$ $ or $100 \cos 36000t°.$

Wave diagrams. It has been shown that it is a simple matter to plot a curve connecting i and t when expressions such as those used in the last paragraph are given, and, instead of giving the trigonometrical expression to represent an alternating current, we may give the curve plotted from that expression thus producing what may be referred to as a wave diagram.

From such a diagram we may see at a glance the maximum value, the frequency may be deduced by observing the time taken to complete a cycle, and the relative phases (if several quantities are indicated) by observing the positions where the different curves cut the zero line when rising positively. The wave form can also be accurately represented when necessary but it is not suggested that in each wave diagram great care need be taken to accurately represent the true shape of the wave. If care be taken to show the maximum values and the relative phases and the curves be neatly sketched in, such wave diagrams are frequently useful in indicating the phenomena taking place in alternating current circuits.

Example. Construct a wave diagram to indicate the quantities dealt with in the last example.

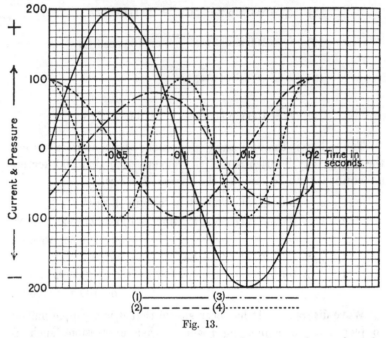

Fig. 13.

Vector diagrams. The two methods given above for the representation of alternating quantities have really been derived from the consideration of a rotating line, and the third method consists in the use of the line itself. The line may be drawn of such a length that, to a suitable scale, it represents the maximum value of the quantity concerned (as a matter of fact the length is more often taken in practice to represent the R.M.S. value of the quantity, see page 20) and if it is desired to represent several quantities the angles between the lines may be used to denote the phase differences. Such lines are termed rotating vectors and a collection of them is termed a vector diagram. The length of the projection of the vector, taken on a line drawn at right angles

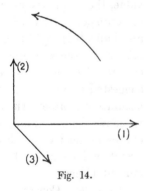

Fig. 14.

to the zero position of the vector, will give the instantaneous value of the quantity concerned*. When we represent currents or pressures by a vector diagram we assume the wave forms to be sinusoidal, and the diagram, in itself, gives no information in regard to the frequency but, as indicated before, this is not usually of great importance. In conclusion we may state that a vector diagram is the simplest and most commonly used way of representing alternating quantities; when one is used care should be taken to represent the lines as moving in a counter-clockwise direction (a matter of convention) and to place an arrow-head on the moving end of the line.

Example. Indicate by a vector diagram the quantities 1, 2 and 3, dealt with in the foregoing examples. See Fig. 14.

General Properties of Alternating Currents.

Chemical effect. Students with a knowledge of the chemical effect of direct current will realise that the substitution of alternating current will cause considerable modification in the results on account of the fact that the effects obtained during one half cycle may be nullified by the effects obtained during the next half cycle. Thus, when a solution of copper sulphate is electrolysed, copper electrodes being used, the electrode which is the anode during one half cycle has copper taken from it while the one which acts as the cathode has copper deposited on it from the solution. During the next half cycle the electrode which has previously acted as the anode becomes the cathode and thus the metal taken off during the previous half cycle is redeposited and *vice versâ*, the resultant electrolytic effect during the complete cycle being zero. In some cases a resultant effect may be obtained as, for instance, when acidulated water is electrolysed using platinum electrodes; in this case it would seem that the products of electrolysis (hydrogen and oxygen) have time to get away from their respective electrodes before the current reverses and so a considerable amount of gas is liberated when an alternating current is passed. The amount of liquid split up is not so great with alternating as with direct current, and the amount is less the higher the frequency.

---

* If the length of the vector represents the maximum value.

Results obtained experimentally are shown in the following table, equal currents and times being used in each case.

Type of current	Quantity of gas liberated
Direct current	47·9 c.c.
Alternating current at 25 cycles per sec.	7·2 „
„ „ 50 „ „	3·3 „
„ „ 75 „ „	0·9 „

The electrolytic effect is of considerable importance in connection with the use of uninsulated returns for traction systems; with direct current the maximum drop allowed in such a return is 7 volts on account of the risk of corrosion to gas and water pipes; with alternating current, since the electrolytic effect is smaller, there is the possibility of the use of larger pressure drops.

Some experiments by Professor Haldane Gee would seem to show that the corrosion of iron buried in the ground is, with alternating current, not likely to exceed ·5 % of that occurring with an equivalent direct current and in the case of lead not more than ·1 %*.

Advantage may be taken of the chemical effect of an alternating current in order to determine its frequency by placing a piece of blotting paper, saturated with a solution of potassium ferricyanide, on a drum revolving at a known rate. Two contacts, made of iron wire, are allowed to press side by side on the paper the distance between them being about half an inch and these contacts are connected to the alternator, the current being limited by the insertion of a lamp. Each wire in turn becomes the anode and causes a blue mark on the paper, the cathode not causing any noticeable effect, and if the number of marks occurring in a measured distance be noted the frequency of the current may be found by an obvious calculation.

Magnetic effect. An alternating current, when passed through a straight conductor or round a solenoid, exerts a magnetic effect on soft iron very similar to that occurring with direct current. The magnetic force exerted on any iron in the vicinity of the solenoid is of course of an alternating nature, and the resulting mechanical force will be of a pulsating nature, varying from zero

* *J. I. E. E.*, Vol. 41.

to a maximum as the current varies from zero to a maximum, but not reversing. The fact that the pull does not reverse is due to the attractive force being co-jointly proportional to the strength of the magnetic force and to the strength of the magnetic field produced within the iron; when the direction of the magnetic force changes so does the direction of the induced magnetic field within the iron, thus leaving the direction of the mechanical force the same as before. The same reasoning indicates that the magnitude of the mechanical force will, at any instant, be proportional to the square of the current (apart from the possibility of saturation of the iron). This pulsating of the mechanical force is likely to cause humming and chattering in single phase alternating current magnets unless special precautions are taken. Since the magnetic field within the core of an alternating current electro-magnet is alternating, it will always be necessary to have the cores laminated in order to minimise the eddy current loss which would give rise to excessive loss of power and heating. In any case there will be some loss due to hysteresis and to eddy currents in the individual laminations.

Again, if metallic formers are used to support the coils, it will

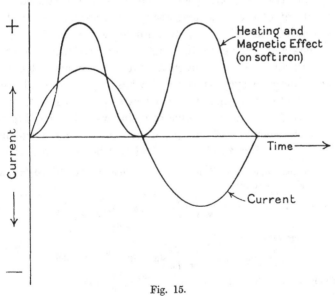

Fig. 15.

be necessary to split them longitudinally in order to prevent the
formation of eddy currents—this device is often seen in the
support for the coils in ammeters and voltmeters of the moving
iron type for use on alternating current circuits.

Heating effect. There is perhaps nothing of a particularly
noteworthy nature in connection with the heating effect of an
alternating current beyond the fact that, since it is at any instant
proportional to the square of the current, it will be of a pulsating
nature and will in fact vary with time much in the same way as the
magnetic effect. As stated before this is likely to cause trouble
due to flickering in metallic filament lamps having fine filaments.

Magnitude of an alternating current. Having noticed the series
of changes of magnitude undergone by an alternating current
during one cycle the point arises as to the best method of expressing
the magnitude of such a quantity. We might, for instance, state
its maximum value or its average value but a little thought will
show that neither is the most convenient way. The power
dissipated by a current when flowing through a resistance is
proportional to i^2, and it at once strikes us as convenient to agree
that an alternating current of X amperes should be such as to
produce an average heating effect equal to that produced by a
direct current of the same magnitude. Thus when we speak of
an alternating current of say 10 amperes we do not mean one
whose maximum value is 10 amperes, or one whose average
value is 10 amperes, but one such that the average value of i^2
during a complete cycle is 100 and whose square root of mean
square value (usually called the root mean square or R.M.S. value)
is 10 amperes. The R.M.S. value of a sine wave can be obtained
by a simple calculation* but generally, and for more irregular
waves, the following graphical method will be convenient.

* Let the maximum value of the current be 100 units, then $i = 100 \sin \theta$, and
the R.M.S. value will be equal to

$$\sqrt{\frac{1}{\pi} \int_0^\pi (100 \sin \theta)^2 . d\theta} = 100 \sqrt{\frac{1}{\pi} \int_0^\pi \sin^2 \theta . d\theta} = 100 \sqrt{\frac{1}{\pi} \int_0^\pi \left(\frac{1 - \cos 2\theta}{2}\right) . d\theta}$$

$$= 100 \sqrt{\tfrac{1}{2}} = 70 \cdot 7,$$

and the average value will be

$$\frac{1}{\pi} \int_0^\pi 100 \sin \theta . d\theta = \frac{100}{\pi} \int_0^\pi \sin \theta . d\theta = \frac{100}{\pi} \times 2 = 63 \cdot 6.$$

and, finally, the form factor for a sine wave will be $\dfrac{70 \cdot 7}{63 \cdot 6} = 1 \cdot 11.$

(1) Draw half a complete wave (this is sufficient in practice since the positive and negative half waves are similar except as regards sign) and divide the axis of time into any convenient number of equal parts (say ten).

(2) Erect and measure the mid-ordinates of each part.

(3) Square each ordinate and obtain the mean square.

(4) Extract the square root and so obtain the square root of mean square value.

It will be obvious that the value of the ratios $\dfrac{\text{Maximum value}}{\text{R.M.S. value}}$

and $\dfrac{\text{R.M.S. value}}{\text{Average value}}$ will depend on the shape of the curve; the latter ratio is known as the *form factor* and is of importance in calculating the pressure given by an alternator or transformer. By drawing curves of various shapes (sinusoidal, flat-topped and peaked) and making suitable calculations, students may investigate the manner in which the shape of the curve influences the form factor.

Example. Determine the R.M.S. and average values of the current whose wave form is given in Fig. 16 and calculate the form factor.

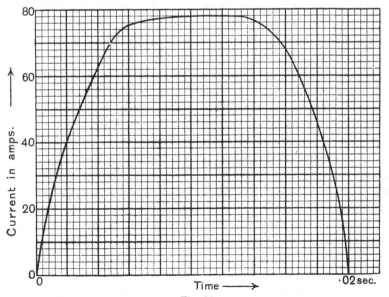

Fig. 16.

Number of mid-ordinate	Length	(Length)²
1	24	576
2	52	2704
3	72	5184
4	76	5776
5	78	6084
6	78	6084
7	78	6084
8	74	5476
9	58	3364
10	30	900
Total	620	42232
Average	62	4223·2
R.M.S. value		65

$$\text{Form factor} = \frac{\text{R.M.S. value}}{\text{Average value}} = \frac{650}{620} = 1 \cdot 048.$$

Addition of Sinusoidal Currents or Pressures.

It often happens that two or more alternating pressures are connected in series and we desire to determine the magnitude and phase of their resultant, or again, several currents of different magnitudes and phases may be tapped off the same conductors and we desire to know the magnitude of the resultant current and its phase. In order to arrive at a suitable method for carrying out the above operation consider the simple case of the addition of two pressures. In the first place it will be fairly obvious that we may determine the *instantaneous* value of the resultant by adding together the corresponding *instantaneous* values of the quantities concerned, and it will also be realised that the maximum value of the resultant will only be equal to the arithmetical sum of the maximum values of the components when the latter are in phase, in every other case it will be less since the maximum values of the components do not occur simultaneously (see Fig. 17). Suppose we wish to determine the sum of the waves shown by the continuous and dotted lines in that figure; these quantities can also be represented by the rotating vectors OA and OB.

Complete the parallelogram of which OA and OB form the sides and draw the diagonal OC. Looking upon the line OC as representing another sinusoidal quantity, we see that it is such that its instantaneous value is always equal to the sum of the instantaneous values of OA and OB or in other words it represents the result of adding the two component waves together. Two important conclusions follow from the above, in the first place we see that the sum of two or more sinusoidal quantities is itself a sinusoidal quantity, and, further, we see that it is possible to add together two or more sinusoidal quantities by a graphical method exactly analogous to that used in the parallelogram of forces.

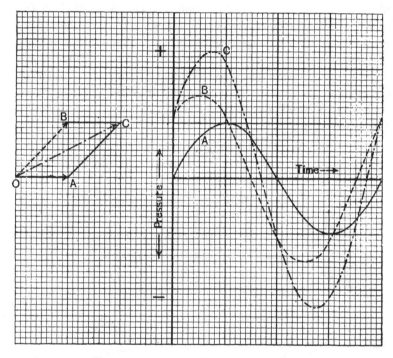

Fig. 17

Example. A pair of mains supplies current to three consumers who take 40, 20 and 60 amperes respectively. If the current taken by the second consumer leads 45°, and the current taken by the third consumer lags 30°, compared with that taken by the first, determine the resultant current flowing along the mains and its phase relative to the pressure.

Having chosen a suitable scale (in this case $\frac{3}{4}''$ represents

50 amperes) it is desirable to first draw a simple vector diagram as shown in Fig. 18 (b), carefully marking the correct angles, lengths and arrow-heads; this procedure will save the possibility of subsequent mistakes *.

Next draw pq (Fig. 18 (c)) parallel to OA making it ·6 inch long and placing on it the correct arrow-head; from the end q of pq draw ql parallel to OB making it ·3 inch long and placing on it the arrow-head, noting that it is circuital with respect to that already placed on pq; finally from l draw lm parallel to OC making it ·9 inch long and again taking care that the arrow-head is circuital with respect to the others. The resultant is then pm and on scaling it is found to be 107·5 amperes. Further, the angle qpm represents the phase difference between the resultant current and the pressure (assuming the pressure to be in phase with OA); it is 8·5° (lagging relative to the current taken by the first consumer).

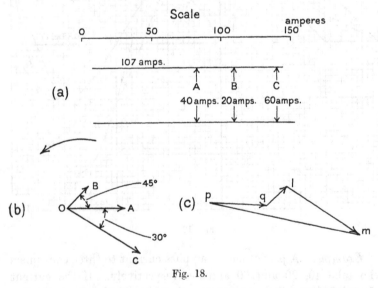

Fig. 18.

If the student does not desire to employ a graphical method he should have no difficulty in inventing for himself a trigonometrical formula (having, of course, greater accuracy) founded

* The lengths may represent either maximum or R.M.S. values, in this case the latter are indicated.

upon the vector diagram for the case in question. Thus in the above example a trigonometrical method may be employed by splitting each of the second and third vectors into two components, one along the line of the first vector and the other at right angles to that direction.

Not only is it possible to combine two components to form a single resultant, but the reverse process may be carried out and a single rotating vector resolved into two components, the single vector in any case being the diagonal of the parallelogram of which the components are the sides. It is interesting to note that while two components can have but a single resultant, a single resultant may have any number of pairs of components.

Example. Solve the last problem by a trigonometrical method.

The component in phase with the first vector will, in each case, be equal to the current multiplied by the cosine of the angle between the current in question and the first current, and the component at right angles will be found by substituting the sine of the angle instead of the cosine (care must be taken in this case to obtain the correct sign). Splitting up each component in this way we obtain the following results:

Current	Component in phase with A	Component at right angles to A
A	40 amperes	0 amperes
B	$20 \times \cos 45° = 14 \cdot 14$ amperes	$-20 \times \sin 45° = -14 \cdot 14$ amperes
C	$60 \times \cos 30° = 51 \cdot 96$ amperes	$+60 \times \sin 30° = +30 \cdot 0$ amperes
Total	$106 \cdot 1$ amperes	$+15 \cdot 86$ amperes

Combining the two totals we get the resultant current to be $\sqrt{(106 \cdot 1)^2 + (15 \cdot 86)^2} = 107 \cdot 5$ amperes; further it lags behind A by an angle ϕ such that

$$\tan \phi = \frac{\text{Component at right angles to } A}{\text{Component in phase with } A} = \frac{15 \cdot 86}{106 \cdot 1} = \cdot 149,$$

whence $\phi = 8 \cdot 5°.$

EXAMPLES

1. A rectangular coil of wire has 20 turns and is 40 cms. long by 20 cms. broad. If it rotates about its long diameter at a rate of 50 revolutions per second in a magnetic field whose density is 100 lines per square cm., calculate the maximum value of the pressure produced. Plot the curve connecting pressure with time.   *Answer.*   5·024 volts.

2. Determine the instantaneous values of the pressures represented by the equation $e = 200 \sin 314t$ when $t = \cdot 001$ and $\cdot 017$ second respectively.

          Answer.   61·8 volts; -162 volts.

3. A current has a maximum value of 80 amperes. If its frequency is 50 cycles per second and it passes through its maximum positive value when $t = 0$, calculate the instantaneous value when $t = \cdot 002$ second.

          Answer.   64·7 amps.

4. Write down all the information you can extract from the vector diagram in Fig. 19 if one inch represents 100 amperes.

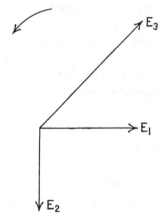

Fig. 19.

5. Express the information gathered from the following formulae in the form of a vector diagram.

    (*a*) $e = 100 \sin 314t$;   (*b*) $e = 200 \cos 314t$;   (*c*) $i = 80 \sin \left(314t - \dfrac{\pi}{4} \right)$.

6. Write down all the information you can extract from the wave diagrams given in Fig. 20.

7. Represent in each of the ways mentioned in the text the following alternating quantities:

    (*a*)   A pressure whose maximum value is 200 volts, frequency 50 cycles per second and which is passing through its zero value, increasing positively, when $t = 0$.

(b) A pressure whose maximum value is 130 volts, frequency 50 cycles per second and which is leading (a) by 30°.

(c) A pressure whose maximum value is 250 volts, frequency 50 cycles per second and which lags behind (a) by $\frac{\pi}{4}$ radian.

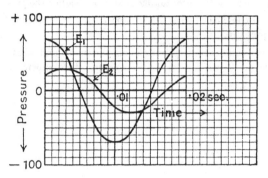

Fig. 20.

8. Determine the R.M.S. value and form factor for:

(a) A sine wave whose maximum value is 200 amperes,

(b) The peaked wave shown in Fig. 21.

Answer. 141·4 volts, 1·111; 31·6 volts, 1·36.

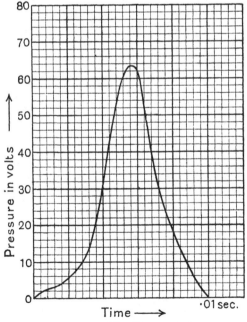

Fig. 21.

9. What shape of wave form has a form factor equal to unity?

Answer. A rectangular wave.

10. The following currents are taken from a pair of A.C. mains:

(*a*) Lamps taking 12 amperes in phase with the pressure.

(*b*) A synchronous motor taking 24 amperes leading the pressure by 40°.

(*c*) A single phase induction motor taking 15 amperes lagging 50° behind the pressure.

What is the total current passing along the mains and what is its phase relative to the pressure? *Answer.* 40·3 amperes, 5·5° leading.

CHAPTER II

INDUCTANCE

In dealing with problems concerning the flow of a steady direct current in a circuit the only source of opposition to the current is the resistance of the circuit, but in many cases when dealing with an unsteady current there is a further source of opposition in the form of induced pressure.

Consider a circuit which comprises a coil of several turns of wire provided with an iron core; when current passes round such a circuit magnetic lines will pass through the core and link with the turns of wire wrapped round it. Suppose the current rises, the magnetising force on the core will increase and more magnetic lines will be generated. Now we regard the lines as springing out from the centre of the core and hence, when they appear or disappear, they must cut the several turns of the conductor thereby producing induced pressures and, in accordance with the law of Lenz, these induced pressures will always be in such a direction as to oppose the change in current strength which is producing them. Thus when the current is rising the induced pressure will be a back pressure tending to oppose the rise of current, and when the current is falling the induced pressure will be a forward one tending to maintain the flow of current.

When dealing with direct currents in such circuits these induced pressures will only be in evidence for the occasional periods when the current is rising or falling, but when dealing with alternating currents (which are continually rising or falling) their presence is much more important and very materially influences the current flow in the circuits in question. Any circuit in which such induced pressures are produced when current

rises or falls is termed an inductive circuit or is said to possess
inductance, and it will be seen that the criterion as to whether
a circuit is or is not inductive is the presence or absence of
magnetic lines linking with the circuit when current is passing
round the circuit.

Strictly speaking it is difficult to imagine a circuit which is
absolutely non-inductive but a circuit is looked upon as non-
inductive, for ordinary purposes, when the linkages between the
turns and magnetic lines are few in number; thus banks of lamps
and water resistances are commonly looked upon as non-inductive
loads so far as what may be called heavy electrical engineering
purposes are concerned.   On the contrary, coils of wire with iron
cores, as field coils and armature coils, are highly inductive.   If
it is desired to construct an electrical circuit whose inductance
shall be as small as possible the disposition of the turns should
be such that the magnetic effect of one part of the conductor is
neutralised by the magnetic effect of another part.   Thus if the
wire is doubled at the centre and the doubled wire thus formed
used to wind the coil, the two strands always being kept near to

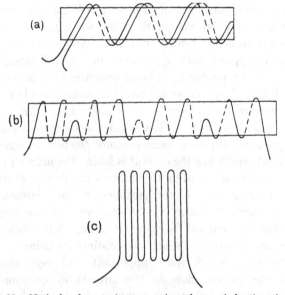

Fig. 22.   Methods of arranging approximately non-inductive circuits.

each other, this end is largely attained; or again, the wire may be wound on a former, first a few turns in one direction and then an equal number of turns in the opposite direction.

Such methods of rendering circuits less inductive than would otherwise be the case are commonly used in constructing the series resistances for the pressure circuits of voltmeters, watt-meters and supply meters.

In constructing diagrams of alternating current circuits it is necessary to differentiate between circuits which are non-inductive and those which are inductive (these may, in addition, be resistive) and this end is attained as indicated in Fig. 23.

Entering next into a few considerations concerning the magnitude of the induced pressure, we see that this will depend upon the rate at which magnetic lines are cut by the turns of the circuit.

Non-Inductive circuit

Inductive circuit

Fig. 23.

Thus if one line is cut per second the induced pressure will be one absolute unit, and if 10^8 lines are cut per second the induced pressure will be one volt. Further, it is easy to see that the rate of cutting magnetic lines will depend upon the rate at which current is altering in the circuit, and also upon some function of the circuit itself which will be a measure of the amount of self-induction inherent in the circuit.

It will be convenient at this stage to grasp the idea of linkages; by this term is meant the product of the number of turns of the solenoid into the number of magnetic lines threading through those turns, thus if 400 turns are threaded by 1000 lines the linkages would be 400,000; the rate at which magnetic lines change in circuit, when multiplied by the turns, gives the rate of change of linkage in the circuit.

If we have a circuit, whose core is air or other non-magnetic material, in which N linkages are produced when one ampere flows round it, then twice as many magnetic lines and consequently $2N$ linkages will be produced when two amperes are passed. In this circuit, when current alters, the rate of change of linkage and consequently the magnitude of the induced pressure will depend co-jointly upon the number of linkages per ampere and upon the rate of change of current, and consequently we may use the number of linkages produced in a circuit when traversed by some standard current as a measure of the amount of self-induction (or inductance) possessed by the circuit.

Absolute unit of self-induction. A circuit is said to have one absolute unit of self-induction when its arrangements are such as to cause the production of one linkage when one absolute unit of current flows round the circuit.

If current rises in this circuit at unit rate (one absolute unit of current per second) the rate of change of linkage will be one per second and the induced pressure will be one absolute unit. This idea may be used as an alternative method of defining the absolute unit of self-induction.

Practical unit of self-induction. A circuit is said to have one practical unit of self-induction when its arrangements are such that 10^8 linkages are produced when one ampere flows round the circuit.

If in this circuit current rises at unit rate (in this case one ampere per second) the induced pressure will be one volt.

The practical unit of self-induction is termed the henry and the amount of self-induction in a circuit, expressed either in terms of the absolute or practical unit, is referred to as the coefficient of self-induction or more usually as the inductance of the circuit. A little thought will show that one henry is equivalent to 10^9 absolute units.

Calculation of the inductance of a circuit. The inductance of any circuit may be calculated for which it is possible to determine the number of magnetic lines linking with the circuit when one ampere is passing round it, and the following examples will serve to show the method to be used. It should be noted that when dealing with coils having cores of magnetic material the inductance will depend very considerably upon the current strength (*i.e.* upon the degree of saturation of the iron), but this difficulty can always be dealt with by determining the linkages for the current in question and then dividing by the current in order to determine the linkages per ampere. If a coil has a core of non-magnetic material the inductance has the same value for all currents, since the number of lines, and consequently the linkages, is proportional to the current. When the core is of magnetic material (and especially when the magnetic circuit is completely composed of iron) a few moments study of the *B—H* curve for the material will show that at first, as we increase the magnetising current, we increase the flux density more than proportionally, hence the linkages per ampere rise giving gradually increasing values of the inductance; as the iron approaches saturation the linkages per ampere attain a maximum value and then continually diminish for increasing currents giving a corresponding diminution of the inductance as indicated in Fig. 24.

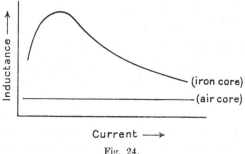

Fig. 24.

Example. A coil of wire, 20 cms. long, is composed of 2000 turns uniformly distributed; if the mean diameter of the turns is 2 cms., determine the inductance of the coil (assume that all the lines generated link with all the turns and that the core is air).

$$H = 1 \cdot 256 \, \frac{IN}{l} = \frac{1 \cdot 256 \times 1 \times 2000}{20} = 125 \cdot 6 \text{ gausses}$$

$$\therefore \ B = 125 \cdot 6 \text{ lines per sq. cm.}$$

Total flux $= 125 \cdot 6 \times \cdot 7854 \times 2^2$ lines.

Linkages per ampere $= 125 \cdot 6 \times \cdot 7854 \times 2^2 \times 2000$.

$$\text{Inductance} = \frac{125 \cdot 6 \times \cdot 7854 \times 2^2 \times 2000}{100000000} = \cdot 0079 \text{ henry.}$$

Example. Calculate the approximate inductance of a coil, consisting of 500 turns of insulated copper wire, wrapped upon a magnetic circuit composed of wrought iron having an air gap 35 sq. cms. in area and ·5 cm. long.

In this example though no particulars of the iron part of the circuit are given yet the approximate inductance can be calculated by assuming that the whole of the reluctance of the magnetic circuit is concentrated in the air gap; on this assumption the number of magnetic lines and consequently the inductance will be rather too high.

$$\text{Magnetic force on the gap} = \frac{\cdot 4\pi I N}{l} = \frac{1 \cdot 256 \times 1 \times 500}{\cdot 5}$$
$$= 1256 \text{ gausses.}$$

Flux density on the gap $= 1256$ lines per sq. cm.

Total flux $= 1256 \times 35$ lines (neglecting fringing).

Linkages $= 1256 \times 35 \times 500$.

$$\text{Inductance} = \frac{1256 \times 35 \times 500}{100000000} = \cdot 22 \text{ henry.}$$

An important application of the above method lies in the calculation of the inductance of armature coils of either alternating or direct current machines. Here we have a case of a coil which is for a part of its length free in air (*i.e.* at the ends), and for a part of its length buried in iron (in the armature core). The most convenient way of working is to make use of constants (determined experimentally) as to the number of magnetic lines produced per ampere per inch length; these may usually be taken as follows:

One ampere flowing through a single wire in air produces two magnetic lines per inch run.

One ampere flowing through a single wire in an open slot situated in iron produces fifteen lines per inch run. This latter number is very variable and will depend much upon the slot opening, the extreme limits being ten and thirty lines per inch run.

Example. Calculate the inductance of the alternator coil whose dimensions are shown in the sketch and which has 20 turns.

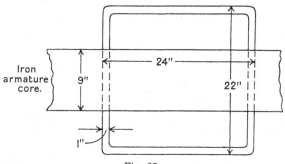

Fig. 25.

Length of wire per turn imbedded in iron = 18″.

Lines per ampere turn due to above = 18 × 15 = 270.

Length of wire per turn in air = 70″.

Lines per ampere turn due to air portion = 70 × 2 = 140.

Total magnetic lines per ampere turn = 140 + 270 = 410.

       ,,   ,,   ,,   coil = 410 × 20 = 8200.

Linkages per ampere coil = 8200 × 20 = 164000.

Inductance $= \dfrac{164000}{10^8} = \cdot00164$ henry.

Example. An anchor ring of circular cross-section is composed of wrought iron and has internal and external diameters of 20 and 24 cms. respectively. If it is uniformly over-wound with 400 turns, calculate the inductance when one ampere and ten amperes respectively are passing round the coil.

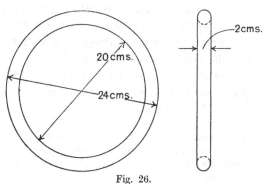

Fig. 26.

(a) Magnetising force exerted by the coil $= \dfrac{1 \cdot 256 \times 1 \times 400}{22 \times 3 \cdot 14}$

$$= 7 \cdot 26 \text{ gausses.}$$

Corresponding flux density (from a typical curve) = 11300 lines per sq. cm.

Total flux = $11300 \times 3 \cdot 14 \times 1^2 = 35500$ lines.

Linkages per ampere = $35500 \times 400 = 14200000$.

Inductance $= \dfrac{14200000}{100000000} = \cdot 142$ henry.

(b) Magnetising force exerted by the coil $= \dfrac{1 \cdot 256 \times 10 \times 400}{22 \times 3 \cdot 14}$

$$= 72 \cdot 6 \text{ gausses.}$$

Corresponding flux density (from a typical curve) = 16500 lines per sq. cm.

Total flux = $16500 \times 3 \cdot 14 \times 1^2 = 51900$ lines.

Linkages per 10 amperes = $51900 \times 400 = 20760000$.

Linkages per 1 ampere = 2076000.

Inductance $= \dfrac{2076000}{100000000} = \cdot 02076$ henry.

This example illustrates the dependence, in certain cases, of the inductance upon the current strength.

From what has been said concerning the definition of the henry the student will realise that the induced pressure in any circuit, due to self-induction, is equal to LI_1, where L is the inductance of the coil and I_1 is the rate of rise or fall of current, and, further, the direction of this induced pressure will be such as to oppose the alteration in current strength which is the cause of its existence. If this induced pressure is taken into account then Ohm's Law may be applied to a circuit at any instant no matter how quickly the current may be varying or how inductive the circuit may be; the only precaution we need take is to use the effective pressure (which will be the algebraical sum of the applied and induced pressures) instead of the applied pressure as would be the case with steady currents*.

---

* Pressures may be produced in a coil not only due to self-induction, as above, but also by the current variation in a neighbouring coil, if, as often happens, the two coils have mutual induction (i.e. current variation in one coil produces a change of flux in the adjacent coil).

$$\text{Current} = I = \frac{\text{Effective pressure}}{\text{Resistance}} = \frac{E - Ll_1}{R} \text{ or } E = IR + I_1 L.$$

In this formula, if the current is rising, the induced pressure will be a back pressure an b I_1 must be taken as positive; if the current is falling I_1 must be taken as negative and the induced pressure will tend to prevent the fall of current which is producing it. Using the above formula many interesting problems may be dealt with throwing instructive light on the rise and fall of current in inductive circuits.

Example. An alternating current (of sinusoidal wave form) has a maximum value of 30 amperes and a frequency of 50 cycles per second. If the resistance of the circuit is 10 ohms and the inductance ·02 henry, calculate (from first principles) the curve of applied pressure necessary to drive the current through the circuit.

Now the applied pressure at any instant is made up of two parts: (1) the applied pressure necessary on account of the induced pressure, and (2) the applied pressure necessary on account of the resistance of the circuit. So long as we deal with instantaneous values we may add these two components algebraically. The applied pressure on account of resistance at any instant can be found by the formula iR. To determine the induced pressure at any instant we must find the rate of alteration of current (either by a mathematical method or a graphical method involving the drawing of tangents to the curve at the several points taken) and if this value is multiplied by the inductance we shall obtain the induced pressure. The necessary applied pressure will then be of the same magnitude but of the opposite sign. Thus when $t = \cdot0025$ second the instantaneous value of the current is 21·2 amperes from the current curve (which may be plotted as shown on page 5) and therefore the applied pressure required on account of resistance is 21·2 × 10 or 212 volts. On very carefully drawing the tangent to the curve at this point the current is seen to be rising at the rate of 21·2 amperes in ·00318 second or 6670 amperes per second*.

---

* But poor accuracy will result if the rate of rise of current is determined graphically and the numbers in the table have been determined by the calculus method.

If $i = 30 \times \sin \omega t$ then the rate of rise of current $= 30\omega \cos \omega t$, where $\omega = 2\pi f$ and $\omega t = 45°$ for the point under consideration.

Rate of rise of current $= 30 \times 314 \times \cdot707 = 6650$ amperes per second.

The induced pressure on account of self-induction is therefore
− 6650 × ·02 or − 133 volts, and the applied pressure is + 133
volts. The total applied pressure is, at this instant,

$$212 + 133 = 345 \text{ volts.}$$

A complete set of results is shown in the table and the curves are
plotted in Fig. 27.

Time (from start)	Current	Rate of current rise	Pressure to drive current through resistance	Induced pressure due to inductance	Applied pressure necessary owing to inductance	Total applied pressure
Seconds	Amperes	Amperes per sec.	Volts	Volts	Volts	Volts
·0	0	+9420	0	− 188	+188	+188
·0025	+21·2	+6650	+212	− 133	+133	+345
·005	+30·0	0	+300	0	0	+300
·0075	+21·2	− 6650	+212	+133	− 133	+ 79
·01	0	− 9420	0	+188	− 188	− 188
·0125	− 21·2	− 6650	− 212	+133	− 133	− 345
·015	− 30·0	0	− 300	0	0	− 300
·0175	− 21·2	+6650	− 212	− 133	+133	− 79
·0200	0	+9420	0	− 188	+188	+188

EXAMPLES

1. An anchor ring of circular cross-section has internal and external
diameters of 16 and 20 cms. respectively. What will be the inductance of
the circuit if the core is composed of iron having a permeability of 600 and is
over-wound with 400 turns?   *Answer.* ·067 henry.

2. A magnetic circuit has in it an air gap 12 sq. cms. in area and 1 cm.
long. If it is over-wound with 500 turns, calculate the approximate inductance
thereby produced.   *Answer.* ·0377 henry.

3. An alternator has 12 coils per phase and 4 turns per coil. If each coil
is of the shape shown in Fig. 25, calculate the inductance per phase. Assume
that one wire produces 15 lines per inch per ampere when in the slot.
    Answer. ·000788 henry.

4. An anchor ring of circular cross-section, having internal and external
diameters of 20 and 24 cms. respectively, is composed of wrought iron. If it
is over-wound with 800 turns, calculate the inductance for currents of ·2, 1, 2,
5, 10 and 20 amperes respectively. Plot a curve connecting current and
inductance. (A typical B—H curve should be used and the answers given will
only be approximate.)   *Answer.* ·95, ·34, ·19, ·081, ·041, ·021 henry.

5. If the current indicated by the expression $i = 24 \times \sin(314t)$ is flowing through a purely inductive circuit whose inductance is ·02 henry, calculate the value of the induced pressure when $t = ·001$, ·00175 and ·013 second respectively. *Answer.* -143, $-128·5$, $+89$ volts.

6. What will be the value of the applied pressures necessary at the times stated in the last example if, in addition, the circuit has a resistance of 10 ohms? *Answer.* $+217$, $+254$, -283 volts.

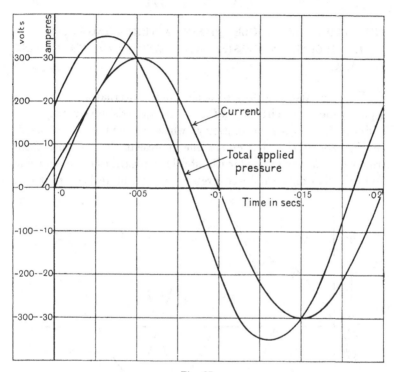

Fig. 27.

CHAPTER III

THE FLOW OF SINGLE PHASE ALTERNATING CURRENTS IN CIRCUITS POSSESSING RESISTANCE, INDUCTANCE AND CAPACITY

Consider first the flow of a sinusoidal alternating current in a purely inductive circuit (that is one whose resistance is negligible and which possesses no capacity) whose inductance is constant. In Fig. 28 let the continuous line represent the current, then, from what has been stated in Chapter II in regard to the magnitude and direction of the induced pressure, it will be realised that this

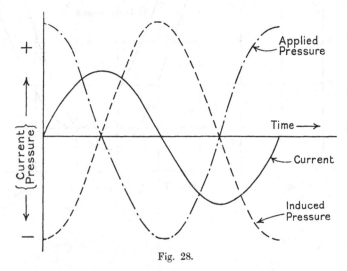

Fig. 28.

is represented (speaking generally) by the dotted line. Notice particularly that when the slope of the current curve is small the rate of change of current, and hence the induced pressure, is also small. The precise shape of the induced pressure curve is perhaps rather more difficult to obtain but with the aid of the calculus it can readily be shown to be another sinusoidal curve displaced 90° relative to the current curve*.

Students unable to follow the footnote may realise the exact shape of the pressure curve by remembering that the current curve may be looked upon as produced by plotting the length of the projection of a rotating line OA (whose length represents I_m and which passes through the position OA when $t = 0$) on the line XY, against time. When the line, see Fig. 29, is passing through the position OA the circumferential speed of the point A, and consequently the rate of rise of current, is $2\pi f I_m$ or ωI_m; but when the line is passing through the position OB (giving $i = I_m \sin \omega t$) though the circumferential velocity is still ωI_m (represented by BD) the rate at which the extremity of the line, and consequently the current, is rising is represented by BC.

Now $BC = BD \cos \omega t = \omega I_m \cos \omega t$ or $\omega I_m \sin (\omega t + 90°)$, and the induced pressure will be $-\omega L I_m \sin (\omega t + 90°)$.

The applied pressure necessary to drive the current round the circuit will, since the resistance is negligible, be equal in magnitude but opposite in phase to the induced pressure and will therefore be represented by $\omega L I_m \sin (\omega t + 90°)$ and by the chain line in Fig. 28†.

To drive a current whose maximum value is I_m amperes with a frequency of f cycles per second through a circuit whose inductance is L henries requires a pressure whose maximum value is $2\pi f L I_m$ volts.

$$E_m = 2\pi f L I_m = \omega L I_m$$

or

$$I_m = \frac{E_m}{\omega L}.$$

* Now $i = I_m \sin \omega t$ and the induced pressure is equal to $-L\dfrac{dI}{dt}$ (the rate of change of current being taken as positive when the current is rising)

$$= -L\frac{d\,(I_m \sin \omega t)}{dt} = -L\omega I_m \cos \omega t = -L\omega I_m \sin (\omega t + 90°)$$

and the applied pressure will therefore be $L\omega I_m \sin (\omega t + 90°)$.

† Students will probably find it difficult to realise that the applied and induced pressures just balance each other; they should think of the induced pressure only existing in virtue of the current flowing and that this automatically takes up such a value that the pressures balance. We have here an electrical case equivalent to Newton's third law of motion.

Further, since both I and E are sinusoidal, their R.M.S. values will each be $\frac{1}{\sqrt{2}}$ of their maximum values, and the above relationship will still be true if we substitute R.M.S. values for those used above or $I = \frac{E}{\omega L}$.

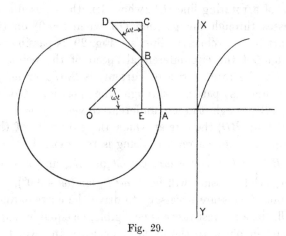

Fig. 29.

The term ωL is known as the *reactance* of the circuit and, though it is not of the same nature as ohmic resistance, yet it limits the current in much the same way as the latter and hence is often expressed as so many *apparent ohms*. A vector diagram showing the phase relations in the above circuit is given in Fig. 30; notice particularly that the current lags 90° (a quarter of a period) behind the applied pressure.

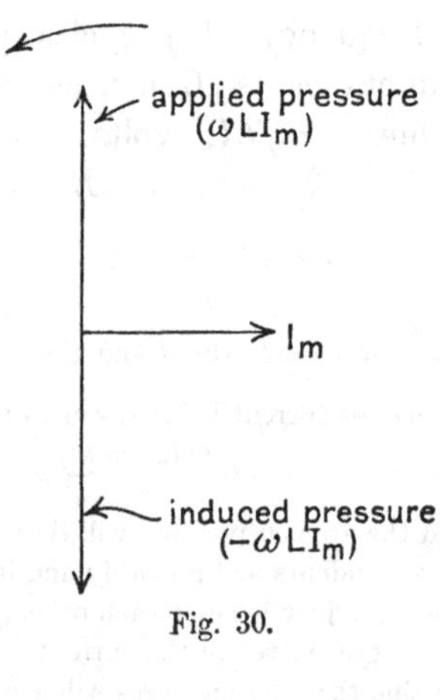

applied pressure
$(\omega L I_m)$

I_m

induced pressure
$(-\omega L I_m)$

Fig. 30.

Flow of current in circuits containing resistance and inductance. In practice we seldom meet with purely inductive circuits, resistance usually being present to an appreciable extent.

Let us consider the flow of a sinusoidal current of maximum

value I_m at f cycles per second through a circuit whose inductance is L henries and resistance R ohms. In this case we can conveniently divide the applied pressure into two components, that necessary to drive the current through the reactance, and that necessary to drive the current through the resistance, respectively. In the vector diagram shown in Fig. 31, if I_m is the maximum value of the current, the applied pressure necessary on account of the resistance will be in phase with the current and its maximum value will be RI_m volts.

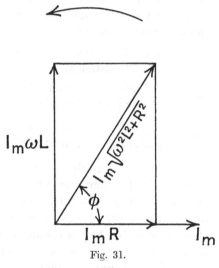

Fig. 31.

The pressure necessary to drive the current through the reactance will, as shown above, have a maximum value of ωLI_m volts and will be 90° in advance of the current. The total applied pressure will be the resultant of the components and will have a maximum value of $I_m\sqrt{\omega^2L^2 + R^2}$ volts,

or
$$E_m = I_m\sqrt{\omega^2L^2 + R^2},$$

$$\therefore \quad I_m = \frac{E_m}{\sqrt{\omega^2L^2 + R^2}}.$$

Further, for the same reasons as given above, we may substitute the R.M.S. values for the maximum values in the above expression and so get

$$I = \frac{E}{\sqrt{\omega^2L^2 + R^2}}.$$

The denominator is termed the impedance of the circuit and, as in the case of reactance, is expressed as so many apparent ohms. The numerical relationship between impedance, reactance and resistance is

$$\text{Impedance}^2 = \text{Resistance}^2 + \text{Reactance}^2.$$

Angle of lag of current behind pressure in the above case. This is represented by the angle ϕ in the diagram in Fig. 31 and obviously depends upon the relative values of the two components of the applied pressure $\omega L I_m$ and $R I_m$ or, what is equivalent, the relative values of reactance (ωL) and resistance (R).

We have the following relationships:

$$\tan\phi = \frac{\omega L}{R} = \frac{\text{Reactance}}{\text{Resistance}},$$

$$\sin\phi = \frac{\omega L}{\sqrt{\omega^2 L^2 + R^2}} = \frac{\text{Reactance}}{\text{Impedance}},$$

$$\cos\phi = \frac{R}{\sqrt{\omega^2 L^2 + R^2}} = \frac{\text{Resistance}}{\text{Impedance}}.$$

The whole of the formulae given above can be applied to the entire circuit or to any part of a circuit, care being taken that the values of I, E, R and L used in any one application of the formulae all refer to the whole or to the part of the circuit in question.

Example. A resistance of 10 ohms is connected in series with an inductive resistance whose inductance is ·04 henry and whose resistance is 2 ohms. If a pressure of 200 volts at 50 cycles per second is applied to the combination, determine the current flowing through the circuit, the pressure across each part and the angle of phase difference between pressure and current for the whole circuit and for each part. In addition, make a vector diagram showing the several quantities concerned.

Dealing first with the entire circuit

$$I = \frac{E}{\sqrt{\omega^2 L^2 + R^2}} = \frac{200}{\sqrt{(6 \cdot 28 \times 50 \times \cdot 04^2) + 12^2}} = \frac{200}{\sqrt{12 \cdot 56^2 + 12^2}}$$

$$= \frac{200}{\sqrt{301 \cdot 7}} = 11 \cdot 5 \text{ amperes,}$$

and $\qquad \tan\phi_1 = \dfrac{\omega L}{R} = \dfrac{12 \cdot 56}{12} = 1 \cdot 047, \therefore \phi_1 = 46°.$

Dealing next with the inductive portion of the circuit

$$E = I \times \sqrt{\omega^2 L^2 + R^2} = 11\cdot5 \times \sqrt{(6\cdot28 \times 50 \times \cdot04)^2 + 2^2} = 11\cdot5\sqrt{161\cdot7}$$
$$= 146 \text{ volts,}$$

and $\qquad \tan\phi_2 = \dfrac{12\cdot56}{2} = 6\cdot28, \quad \therefore \; \phi_2 = 81°.$

Finally, dealing with the non-inductive portion of the circuit,

$$E = IR = 11\cdot5 \times 10 = 115 \text{ volts,}$$

and the current and pressure will be in phase for this part of the circuit.

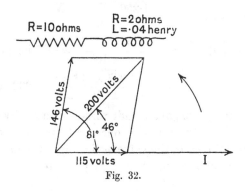

Fig. 32.

Example. An arc lamp requiring 10 amperes at 70 volts is to be placed across mains having a pressure of 110 volts at 50 cycles per second. Calculate the inductance necessary in a choker to be placed in series with the lamp in order to reduce the pressure to the correct amount. Further, if this choker consists of an iron circuit having an air gap 16 sq. cms. in area and ·5 cm. long, determine approximately the number of turns which must be used.

Now the arc lamp may be assumed to be a practically non-inductive circuit, and the choker may be assumed to be purely inductive; the pressures therefore across the two parts of the circuit will be practically at right angles and will be connected by the relationship

(Line pressure)² = (Pressure on lamp)² + (Pressure on choker)²

or (Pressure on choker)² = (Line pressure)² − (Pressure on lamp)²,

$\qquad \therefore$ Pressure on choker $= \sqrt{110^2 - 70^2} = 84\cdot8$ volts.

Reactance of choker $= \dfrac{E}{I} = \dfrac{84\cdot8}{10} = 8\cdot48$ apparent ohms.

$$\text{Inductance of choker} = \frac{\text{Reactance}}{2\pi f} = \frac{8\cdot48}{6\cdot28 \times 50} = \cdot027 \text{ henry.}$$

Let N be the approximate number of turns, then the magnetising force on the gap of the choker (neglecting the reluctance of the remainder of the magnetic circuit) will be

$$\frac{1\cdot256N}{\cdot5} \text{ gausses per ampere,}$$

and

$$\text{gap flux} = \frac{1\cdot256 \,.\, N \,.\, 20}{\cdot5} \text{ lines per ampere.}$$

$$\text{Linkages per ampere} = \frac{1\cdot256 \,.\, N \,.\, 20 \,.\, N}{\cdot5}$$

and

$$\text{inductance} = \frac{1\cdot256 \,.\, N^2 \,.\, 20}{\cdot5 \times 10^8} = \cdot027 \text{ henry.}$$

$$\therefore\ N = \sqrt{\frac{\cdot027 \times 10^8 \times \cdot5}{1\cdot256 \times 20}} = 232 \text{ turns.}$$

Flow of current through a circuit possessing capacity only. The quantity of electricity stored in one conductor of a capacity is co-jointly proportional to the value of the applied pressure between the two conductors forming the capacity and to the magnitude of the capacity.

Consequently, when the value of the applied pressure alters, the quantity of electricity stored in one coating of the condenser alters in proportion and this can only take place by the flow of electricity into or out of the conductor in question. When the charge on one conductor of a condenser varies, the charge on the other conductor alters by practically the same amount but in the opposite direction, and thus a flow of electricity into one coating of a condenser is accompanied by an equal flow out of the other coating; the general effect being, so far as anything outside the condenser is concerned, as if a current had passed through the condenser despite the fact of the presence of the insulating gap *.

* A condenser consists of two coatings or conductors separated by an insulator; when we speak of the capacity of a condenser we really mean the capacity of one coating, and this is measured by the charge necessary to cause unit increase in potential of that coating relative to the other coating.

A condenser is said to have a capacity of one farad when one coulomb is required to raise the potential of the coating one volt. In practice the capacities of condensers are usually expressed in terms of the microfarad which as its name implies. is one millionth part of a farad.

Electrical engineers may meet capacity in the form of artificial condensers but they are far more likely to meet it in connection with cables where we have two conducting cores separated by insulating material thus forming a condenser. Capacity will also exist between each core of the cable and the earth.

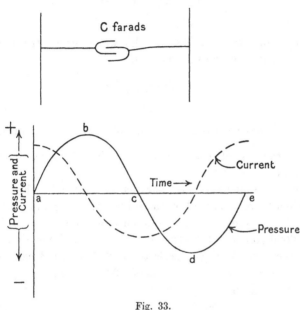

Fig. 33.

Consider the effect of applying an alternating pressure between the two coatings of a condenser whose capacity is C farads. A little thought will show that the current flowing into the condenser at any instant will be

$$C \times \text{rate of change of pressure,}$$

and as the pressure (shown by the continuous line in Fig. 33) passes through its cycle it will be seen that the current will have maximum positive values at a and e, a maximum negative value at c and zero values at b and d.   The general shape of the current curve will be as shown by the dotted line and it is seen to lead the pressure by a quarter of a period.   The exact shape of the current wave may be shown to be a sine curve by a method similar to that used in the case of inductance on page 41, or it may be more briefly demonstrated as in the footnote*.

Now if E_m is the maximum value of the applied pressure the maximum value of the rate of change of pressure will be $2\pi f E_m$ and the maximum current taken by the condenser will be $2\pi f E_m C$,

$$\therefore I_m = 2\pi f E_m C = \frac{E_m}{\dfrac{1}{\omega C}}.$$

Since both waves are sinusoidal we may substitute R.M.S. values for maximum values and write

$$I = \frac{E}{\dfrac{1}{\omega C}}.$$

The term $\dfrac{1}{\omega C}$ has been referred to as the *condensance* of the circuit and, as in the case of reactance and impedance, may be expressed in apparent ohms.

Example. Calculate the charging current (*i.e.* the current taken when there is no load on the cable) for a single phase feeder, which is 10 miles long, when a difference of pressure whose R.M.S. value is 6000 volts at a frequency of 50 cycles per second is applied. Assume the capacity of the cable to be ·2 microfarad per mile.

Total capacity = 2 microfarads = ·000002 farad.

Charging current

$$= \frac{E}{\dfrac{1}{\omega C}} = \frac{6000}{\dfrac{1}{6\cdot28 \times 50 \times \cdot000002}} = 6000 \times 6\cdot28 \times 50 \times \cdot000002$$

$$= 3\cdot77 \text{ amperes}.$$

Of course this is the value of the current taken by the station end of the cable and the value will gradually fall off, since the capacity between the cores of the cable is uniformly distributed along the cable, as the remote end of the cable is reached.

In view of the rapidly increasing use of condensers for general purposes and the further possibility of their introduction in power

---

* At any instant

$$I = C\frac{de}{dt} = C\frac{d\,(E_m \sin \omega t)}{dt} = CE_m\,\omega \cos \omega t = C\omega E_m \sin (\omega t + 90°),$$

or we see that when a sinusoidal pressure is applied to a capacity the resulting current is also sinusoidal and leads the pressure by a quarter of a period.

electrical engineering, for the purpose of improvement of power factor, a brief description of modern forms of condensers may not be out of place. The two chief types in use are:

(1) Foiled paper condensers which are used essentially for low pressure work, up to say 500 volts.

(2) Condensers insulated with glass (really an improvement of the old Leyden Jar) known as Moscicki condensers and which are suitable for pressures of 10,000 volts or even higher.

Foiled paper condensers. The development of this type of condenser has been largely due to Mr G. F. Mansbridge who has dealt with the matter in an excellent paper read before the Institution of Electrical Engineers*.

In preparing these condensers very finely divided tin is deposited on one side of thin paper which is afterwards passed between heated steel rolls under considerable pressure in order to flatten out the particles and convert them into a conducting coherent film. This process sometimes results in unduly large particles of tin being forced through the paper which might cause a short circuit between the coatings of the finished condenser. It is necessary therefore to localise and remove these faults and this is effected by passing the coated paper over rolls so arranged that a pressure of about 100 volts is applied between the coating and the roll over which the uncoated side is passing. When a fault passes the testing point a current flows through the particle of metal which has pierced the paper, and apparently the heat developed is sufficiently intense to vaporise the thin layer of metal in the vicinity thus automatically clearing the fault. The condenser is then made by rolling two sheets of the foiled paper (separated by one or more sheets of plain paper) over a mandril, the area of foiled paper used being dependent on the amount of capacity required.

The condenser is then thoroughly dried in a vacuum oven, impregnated with wax under a suitable pressure and finally pressed, while hot, into any desired shape. Impregnating with wax increases the capacity of the condenser and prevents moisture re-entering the condenser when it cools. When used on alternating current circuits such condensers have been found by Mr Mordey

* *J. I. E. E.*, Vol. 41, p. 535.

to have a very low power factor (certainly not greater than ·01), the small loss which occurs being partly due to resistance loss in the coatings and partly to dielectric hysteresis loss.

If it is desired to use such condensers on high pressure circuits for the improvement of power factor, two methods suggest themselves:

(1) The use of a number of condensers in series.

(2) The use of a special step-down transformer, the condensers being connected on the low pressure side. In this case, in accordance with the well-known action of the transformer, if the current leads the pressure on the low tension side, the primary current will also lead the applied pressure though not perhaps to the same extent (on account of the magnetising current).

Condensers insulated with glass*. These condensers have been developed by Moscicki and consist of glass tubes closed at one end and coated internally with a thin layer of silver, the external coating consisting either of another silver coating or of a conducting solution in which the tubes are partially immersed. The glass tubes are specially shaped at the angles in order to keep down the electrostatic stress at those points. This type of condenser has also been found to have a very low power factor even when used on high pressure circuits.

In conclusion it should be stated that while the first cost of condensers for use on power circuits would be high yet the upkeep and cost of energy wasted would be comparatively very low.

Circuits containing inductance, capacity and resistance in series. Consider the circuit shown in Fig. 34; we may divide the pressure necessary to drive a current of I amperes (we will now deal directly with R.M.S. values) through this into three parts:

(1) That required on account of the resistance; this will have a magnitude of IR volts and will be in phase with the current.

(2) That required on account of the inductance; this will have a magnitude of $I\omega L$ volts and will lead the current by 90°.

(3) That required on account of the condensance; this will have a magnitude of $\dfrac{I}{\omega C}$ volts and will lag 90° behind the current.

* See paper by Mr Mordey, *J. I. E. E.*, Vol. 43, p. 618.

These pressures are shown in the vector diagram and we notice at once that the pressures $I\omega L$ and $\dfrac{I}{\omega C}$ are in direct opposition to each other and may therefore be subtracted arithmetically, giving a resultant pressure on account of reactance and condensance of $\left(I\omega L - \dfrac{I}{\omega C}\right)$ volts.

To get the total applied pressure this must be compounded with IR volts at right angles to itself, the resultant pressure being

$$\sqrt{\left(I\omega L - \frac{1}{\omega C}\right)^2 + I^2 R^2} = I\sqrt{\left(\omega L - \frac{1}{\omega C}\right)^2 + R^2}\text{ volts,}$$

or

$$I = \frac{E}{\sqrt{\left(\omega L - \dfrac{1}{\omega C}\right)^2 + R^2}}.$$

The denominator represents the resultant impedance and, as usual, will be expressed in apparent ohms. The angle of phase difference between current and total pressure is ϕ and it will be seen that

$$\tan\phi = \frac{\omega L - \dfrac{1}{\omega C}}{R}.$$

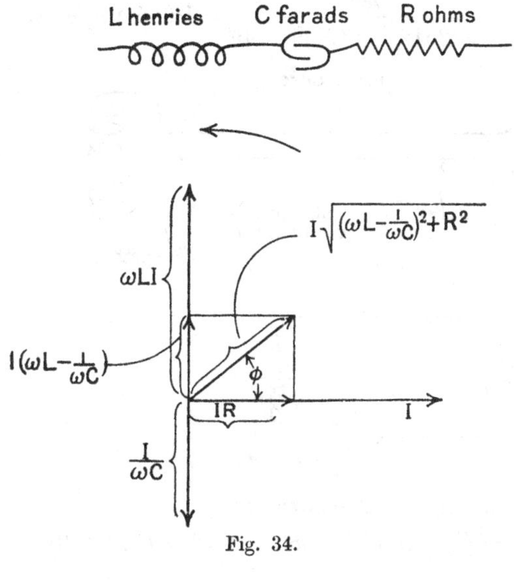

Fig. 34.

If the result is positive a lagging current is indicated, and if negative a leading current.

The above formulae show that as regards limiting the current in a circuit, and as regards producing phase difference between current and pressure, capacity and inductance tend to neutralise each other's effects.

Example. A concentric cable supplies current at a pressure of 2000 volts and a frequency of 100 cycles per second to a transformer. If the end of the transformer connected to the outer conductor of the cable becomes earth connected and a break occurs between this end of the transformer and the conductor, calculate the pressures produced across the transformer and between the outer conductor of the cable and earth.

Capacity between outer core and earth = 1·2 microfarads.

Inductance of transformer = 4 henries.

Resistance of transformer fault to earth = 60 ohms (including resistance of transformer winding, etc.).

Applying the above formula to the complete circuit we have

$$I = \frac{2000}{\sqrt{\left(628 \times 4 - \dfrac{1}{628 \times \cdot0000012}\right)^2 + 60^2}}$$

$$= \frac{2000}{\sqrt{(2512 - 1327)^2 + 60^2}} = 1\cdot686 \text{ amperes.}$$

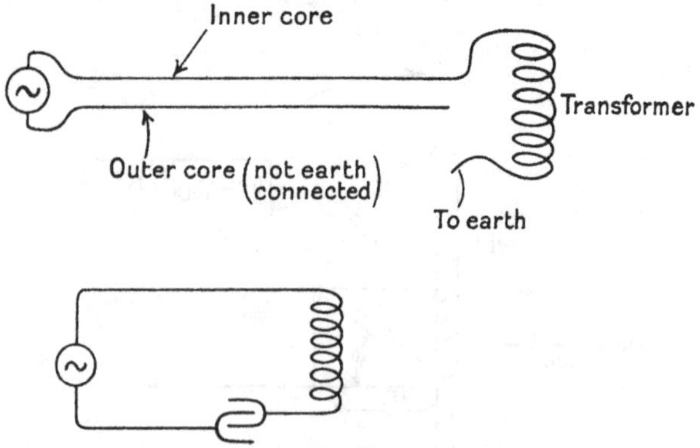

Fig. 35.

Pressure across the transformer

$$= E_t = I\omega L = 1\cdot686 \times 2512 = 4240 \text{ volts.}$$

Pressure across the insulation between the outer core and earth

$$= E_C = \frac{I}{\omega C} = 1 \cdot 686 \times 1327 = 2240 \text{ volts.}$$

In the above example we notice that the pressures across certain parts of the circuit are greater than the pressure across the whole circuit, and, further, the current flowing is greater than the current which would flow were the capacity alone, or the inductance alone, connected across the same pressure. These phenomena are of course due to the partial neutralisation of the effects of the capacity by the effects of the inductance, as indicated by the formula, and are referred to as being due to *electrical resonance*. The physical explanation of these phenomena hardly comes within the scope of this book but we shall get some effect of this kind whenever we have inductance and capacity in series and the effect is most apparent when the capacity and inductance are connected by the relationship $\omega L = \dfrac{1}{\omega C}$, in which case complete neutralisation of these terms takes place and the magnitude of the current flowing depends simply on the resistance of the circuit.

Example. An inductive resistance having an inductance of 2 henry and a resistance of 40 ohms is connected in series with a capacity of 10 microfarads. At what frequency will there be the maximum current through the circuit? If a pressure of 200 volts is applied at this frequency what will be the current flowing, the pressures across the two parts of the circuit and the angles of phase difference between current and pressure for the whole circuit and for each part?

We shall have the maximum current flowing when the condensance and reactance are numerically equal, that is when

$$\omega L = \frac{1}{\omega C} \text{ or } \omega^2 = (2\pi f)^2 = \frac{1}{CL},$$

or when

$$f = \frac{1}{2\pi} \sqrt{\frac{1}{LC}} = \frac{1}{6 \cdot 28} \sqrt{\frac{1}{\cdot 2 \times \cdot 00001}}$$

$$= 112 \cdot 6 \text{ cycles per second.}$$

At this frequency $I = \dfrac{E}{R} = \dfrac{200}{40} = 5$ amperes.

Pressure across the capacity

$$= E_C = \frac{I}{\omega C} = \frac{5}{6 \cdot 28 \times 112 \cdot 6 \times \cdot 00001} = 707 \text{ volts.}$$

Pressure across the inductive resistance

$$= E_L = I \sqrt{\omega^2 L^2 + R^2} = 5 \sqrt{(6 \cdot 28 \times 112 \cdot 6 \times \cdot 2)^2 + 40^2} = 736 \text{ volts.}$$

Angle of phase difference for the complete circuit

$$= \tan^{-1} \frac{\omega L - \dfrac{1}{\omega C}}{R} = \frac{0}{40}, \quad \therefore \ \phi = 0.$$

Angle of phase difference for condensive portion of circuit = 90°, current leading.

Angle of phase difference for inductive portion of circuit

$$= \tan^{-1} \frac{\omega L}{R} = \frac{141 \cdot 4}{40}, \quad \therefore \ \phi_1 = 74°, \text{ the current lagging.}$$

Circuits containing two or more parallel branches. It will be realised that it is useless to determine the current through each branch (by methods already dealt with) and then add them arithmetically, since this method would only give correct results were all the currents in phase and this will not usually be the case. In order to obtain correct results the several components must be added vectorally, that is, we must take into account not only their magnitude but also their relative phase. Let us have a circuit comprising paths in parallel whose impedances are Z_1, Z_2, Z_3, etc., and let the angles of phase difference between currents and pressures in these paths be ϕ_1, ϕ_2, ϕ_3, etc. Then if the applied pressure is E volts the currents through the several paths will be $\dfrac{E}{Z_1}$, $\dfrac{E}{Z_2}$, $\dfrac{E}{Z_3}$, etc. These currents can now be added together and the magnitude and phase of the resultant found precisely as described at the end of Chapter I.

As an alternative we may, instead of adding the currents, add the values of the terms $\dfrac{1}{Z_1}$, $\dfrac{1}{Z_2}$, $\dfrac{1}{Z_3}$, etc. (which may be described as the admittances of the several paths), taking due care to attach the proper phase angle to each. The result will now represent the nett admittance and when multiplied by the pressure will give us the resultant current.

Example. A single phase system having a pressure of 100 volts at 50 cycles per second has connected to it the following loads in parallel:

(1) An arc lamp requiring 10 amperes at 71 volts and having in series with it a suitable choker.

(2) A motor taking 12 amperes leading the pressure by 35°.

(3) A circuit whose inductance is ·02 henry and resistance 10 ohms.

(4) A resistance of 10 ohms.

Determine the resultant current taken from the mains and the phase difference from the applied pressure.

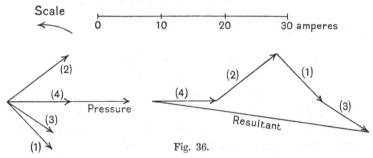

Fig. 36.

Making the necessary calculations we obtain the following particulars concerning the various currents:

(1) 10 amperes lagging 45° behind the applied pressure.

(2) 12 amperes leading 35° in front of the pressure.

(3) 8·45 amperes lagging 32° behind the pressure.

(4) 10 amperes in phase with the pressure.

The vectoral addition of these is carried out in Fig. 36, and the resultant current is found to be 34·6 amperes lagging 8° behind the applied pressure.

The above problem can also be solved by an analytical method if we divide each current into two parts, the first in phase with the current and the second at right angles to it. If we have a current of I amperes differing in phase from the pressure by $\phi°$, the component of the current in phase

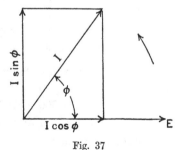

Fig. 37

with the pressure is $I \cos \phi$ and the component of current at right angles to the pressure is $I \sin \phi$ (see Fig. 37).

If we consider a number of currents, then, using the same symbols as before, the total current in phase with the pressure will be the sum of such terms as

$$I_1 \cos \phi_1 + I_2 \cos \phi_2 + I_3 \cos \phi_3, \text{ etc.,}$$

and the total current at right angles to the pressure will be the sum of such terms as $I_1 \sin \phi_1 + I_2 \sin \phi_2 + I_3 \sin \phi_3$, etc., the resultant current then being given by the expression

$$\sqrt{\{(I_1 \sin \phi_1 + I_2 \sin \phi_2 + I_3 \sin \phi_3 + \ldots)^2 + (I_1 \cos \phi_1 + I_2 \cos \phi_2 + I_3 \cos \phi_3 + \ldots)^2\}}.$$

Further, the angle of phase difference between the pressure and current will be given by the expression

$$\tan \phi = \frac{I_1 \sin \phi_1 + I_2 \sin \phi_2 + I_3 \sin \phi_3 + \ldots}{I_1 \cos \phi_1 + I_2 \cos \phi_2 + I_3 \cos \phi_3 + \ldots}.$$

In using these formulae care must be taken in dealing with the signs, thus leading currents may be considered to have positive values of phase difference and then both sines and cosines will be positive, lagging currents will then be considered to have negative values of phase difference in which case the sines will be negative and the cosines positive.

Example. A single phase feeder supplies current to synchronous motor generators and to induction motors, the former take 500 amperes leading the pressure by 45° and the latter 400 amperes lagging behind the pressure by 35°. What is the total current on the line and the phase difference between it and the pressure?

Total current component in phase with the pressure

$$= 500 \cos 45° + 400 \cos (-35°)$$
$$= 500 \times \cdot7071 + 400 \times \cdot8192$$
$$= 681 \text{ amperes.}$$

Total current at right angles to the pressure

$$= 500 \sin 45° + 400 \sin (-35°)$$
$$= 500 \times \cdot7071 + (400 \times -\cdot5736)$$
$$= 124 \text{ amperes.}$$

Total current on feeder $= \sqrt{681^2 + 124^2} = 692$ amperes.

The angle of phase difference is such that $\tan\phi = \dfrac{124}{681} = \cdot182$, whence $\phi = 10°$, the current leading.

Example. An inductive circuit takes 10 amperes (lagging 80°) from mains having a pressure of 200 volts at 50 cycles per second. What capacity should be placed in parallel with the inductive circuit in order that the nett current be in phase with the pressure? What will then be the value of the main current?

Component of above current at right angles to pressure

$= I \sin\phi = 10 \times \sin 80° = 10 \times \cdot9848 = 9\cdot85$ amperes.

Now for a condenser

$$I = \frac{E}{\dfrac{1}{\omega C}}, \therefore C = \frac{I}{E\omega} = \frac{9\cdot848}{6\cdot28 \times 50 \times 200} = \cdot000157 \text{ farad}$$
$$= 157 \text{ microfarads.}$$

When this is placed in parallel with the inductive circuit the current taken from the mains will be

$I \cos\phi = 10 \times \cdot1736 = 1\cdot74$ amperes.

Here we have an instance of the current in each branch of a circuit exceeding the total current taken from the mains; it is another case of the phenomenon known as electric resonance and may be looked upon as being due to the leading and lagging currents neutralising each other as regards the main circuit.

In the case just considered it is not absolutely essential to use a condenser to take the leading current, since a similar current can be obtained in another way and will answer the purpose equally well. Thus, if we arrange that the field current of a synchronous motor (that is an ordinary alternator used as a motor) is in excess of the normal value the machine will take a leading current from the alternating current mains and this may be used to compensate for the lagging current taken by other apparatus on the same line. The line current may thus, if circumstances are favourable, be considerably reduced and the phase difference between pressure and line current reduced to zero or at any rate to considerably less than would otherwise be the case.

Skin Effect.

One curious effect of self-induction is the so-called skin effect which occurs in massive conductors, especially if they possess high

permeability as in the case of iron, and which is also more pronounced at high frequencies.

If we have two precisely similar conductors and through one
we pass a direct and through the other an alternating current, we
should naturally expect to find that a higher pressure was required
in the alternating current case, because even a straight conductor
possesses a certain amount of inductance and, with alternating
current, pressure is required not only on account of resistance but
also on account of inductance.   As a matter of fact if we connected
a wattmeter across each of the wires we should also find that more

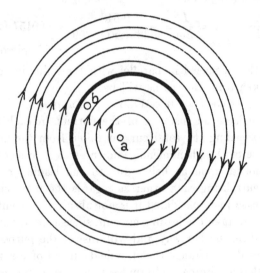

Fig. 38.

power was required in the alternating current case, and this is
not explained by the fact that the wire possesses inductance due
to the magnetic lines set up outside the wire.   In order to investigate the matter imagine the thick circle shown in Fig. 38 to represent
the circular cross-section of a conductor; if an alternating current
is passing through this there will be no magnetic field set up at
the instant the current is passing through its zero value; then,
as the current rises, we imagine magnetic lines of circular shape
to spring out from the centre of the conductor, cutting the material
of the conductor in transit, and setting up therein the usual

induced pressures due to self-induction. At the instant the current has attained its maximum value we can consider the magnetic field as consisting of two parts: the part within the material of the conductor and the part outside the conductor. As the magnetic lines spring out and collapse, due to the rise and fall of current, an element of the conductor at a, situated near the centre of the conductor, will be cut by all the lines of the field. On the contrary an element situated near the side of the conductor, as at b, will only be cut by those lines whose final position is exterior to the conductor. Thus we see that the induced pressures will be greater in elements situated near the centre of the conductor than in those situated near the sides. Now the applied pressure may be regarded as being equal on all elements of the conductor and for any one element the applied pressure will be partly needed on account of resistance and partly on account of inductance, the part needed on account of resistance being proportional to the current flowing in the element.

Since in the centre elements more of the applied pressure will be required on account of self-induction there will be less available to overcome the resistance and consequently the current density will be less near the centre than near the periphery. The current in fact will not distribute itself uniformly over the cross-section, as in the direct current case, but will tend to concentrate in the outer portions of the conductor*.

It is also worthy of note in passing that the phase of the current near the centre will differ from the phase of the current near the periphery. Since the value of the ratio $\dfrac{\omega L}{R}$ is greater near the centre the current will lag to a greater extent in this region than at the periphery.

With copper conductors of sections usually met with in machine and cable work the effect is not great at ordinary frequencies, but at very high frequencies, such as those met with in a lightning discharge or in the oscillatory discharge of a condenser, the skin effect is very pronounced the current being practically confined

---

* Readers should be quite clear that the total applied pressure is not found by adding arithmetically the components necessary on account of resistance and inductance. They must be added vectorally and, since they are at right angles, an increase in one will result in a decrease in the other.

to the surface; the conductivity in such cases is proportional to the periphery rather than to the area of cross-section. Again, if an iron rail is used, as in traction work, as a conductor of alternating current, the effect is very pronounced since the great permeability of the iron gives rise to a strong internal magnetic field causing great disparity between the self-induction of an element near the centre and of an element near the periphery.

EXAMPLES

1. Calculate the current which will flow through a choker of negligible resistance and whose inductance is ·06 henry, when a pressure of 200 volts at a frequency of 50 cycles per second is applied. What relation will exist between current and frequency in this circuit if the other factors remain constant?

Answer. 10·6 amperes; inverse proportionality.

2. Determine the current flowing and the pressure across each part of a circuit consisting of a non-inductive resistance of 4 ohms connected in series with an inductive resistance, whose resistance is 4 ohms and inductance ·04 henry, when a pressure of 100 volts at a frequency of 50 cycles per second is applied. Determine also the angles of phase difference between pressure and current for the whole circuit and for each part.

Answer. 6·74 amperes; 89 volts; 27 volts; 57°; 72°; 0°.

3. An arc lamp requires 10 amperes at a pressure of 55 volts. If the mains pressure is 100 volts at a frequency of 50 cycles per second, calculate the reactance and inductance of a choker suitable for placing in series with the lamp. Calculate the corresponding values if the frequency is 66·6 cycles per second. *Answer.* 8·35 apparent ohms; ·0265 henry.
8·35 apparent ohms; ·0199 henry.

4. The reluctance of the magnetic path of an alternating current electromagnet is ·004 oersted, and it is over-wound with 400 turns. What pressure must be applied to send through the coil a current of 1·3 amperes at a frequency of 50 cycles per second? (Neglect the resistance.) *Answer.* 205 volts.

5. If the resistance of the coil in the above example is 40 ohms, what pressure will then be required? *Answer.* 212 volts.

6. What will be the differences of phase between current and pressure for the complete circuit in example (3)? *Answer.* 57°; 57°

7. A pair of cables 12 miles long have a capacity between them of ·3 m.f. per mile. If the supply pressure is 10,000 volts at a frequency of 50 cycles per second, calculate the charging current taken by the cables.

Answer. 11·3 amperes.

8. Calculate the pressure necessary to drive a current of 10 amperes at a frequency of 50 cycles per second through the following impedances connected in series:

    (a) $R = 85$ ohms, $L = \cdot25$ henry.
    (b) $R = 40$ ohms, $L = \cdot3$ henry.
    (c)   —   — $C = 43$ microfarads.

                                    Answer. 1593 volts.

9. An inductive resistance, for which $R = 40$ ohms and $L = \cdot05$ henry, is connected in series with a capacity of 35 microfarads. If the pressure across the combination is 100 volts at a frequency of 100 cycles per second, determine the current flowing, the pressure across the two parts of the circuit and the angles of phase difference between current and pressure for the whole circuit and for each part.   *Answer.* 2·36 amperes; 120 volts; 107·5 volts; 19·5° (lead); 90° (lead); 38° (lag).

10. Plot a curve, for the circuit dealt with in the last example, connecting the frequency with the current flowing, the other factors remaining constant. Use frequencies from 50 to 200 cycles per second.

11. A choker of ·9 henry inductance is connected in series with a resistance of 50 ohms and a capacity of 23 microfarads. Calculate the frequency for maximum resonance. If the pressure applied is 200 volts at that frequency, calculate the current flowing and the pressure across each of the three parts of the circuit.

Answer. 35 cycles per second; 4 amperes; 793 volts; 200 volts; 793 volts.

12. The following circuits are connected in parallel across a pressure of 220 volts at a frequency of 50 cycles per second. Determine the resultant current flowing and its phase difference from the applied pressure.

    (a) $C = 30$ microfarads.
    (b) $R = 6$ ohms, $L = \cdot02$ henry.
    (c) $R = 20$ ohms.
    (d) $R = 8$ ohms, $L = \cdot01$ henry.

                          Answer. 58 amperes; 26° (current lagging).

13. The load on a certain alternating current feeder is 800 amperes lagging 30° behind the pressure. It is proposed to diminish the line current by connecting in parallel with the load a circuit which will take a leading current. What should be the magnitude of this current, if it is at right angles to the applied pressure, in order to get the best result, and what would then be the current on the line?   *Answer.* 400 amperes; 693 amperes.

14. An inductive circuit takes 8 amperes at 50 cycles per second and lagging 80° behind the pressure. What capacity should be connected in parallel with this circuit, the line pressure being 200 volts, in order that the total current may be in phase with the pressure? What capacity would be necessary were the frequency 100 cycles per second, the other factors remaining the same?   *Answer.* 125·4 m.f.; 62·7 m.f.

CHAPTER IV

POWER IN ALTERNATING CURRENT CIRCUITS

The power at any instant in an alternating current circuit is found by multiplying together the corresponding instantaneous values of current and pressure and, knowing the variations which occur in each of these quantities during a cycle, students will realise that the power in a given circuit will undergo variations both in magnitude and sign.

The question then arises as to what function of the power is understood when we casually refer to the "power" of an alternating current. Since the usefulness of power from the point of view of heating, etc., is proportional to its numerical value, it is clear that what we should endeavour to state in any case is the average power over a cycle or over any number of complete cycles*.

Power in a non-inductive circuit. In this case current and pressure are in phase and, starting our consideration from the instant when both are zero, we see at first power will also be zero and then, as time goes on, will rise but slowly, since both current and pressure are small; as current and pressure attain larger values the power also rises reaching a maximum at the same instant as current and pressure reach their maximum values. As current and pressure decrease, and eventually become zero, power also falls and becomes zero. Current and pressure now simultaneously reverse in sign and thus the sign of power remains positive and during the second half cycle of current and pressure the variations of power will be precisely as in the first half cycle.

* Compare this statement with the reason given for deciding to specify a current or pressure by stating its R.M.S. value; the student should be quite clear as to why the same method is not used in the case of power.

Power at any instant $= p = ie$, but in the circuit under consideration

$$e = iR, \quad \therefore \ p = ie = i^2R,$$

and average value of power over one complete cycle $= P =$ average value of i^2 over one complete cycle $\times R$.

Further the average value of i^2 over a complete cycle $= I^2$,

$\therefore$ average power in circuit in question $= I^2R = IE = \dfrac{E^2}{R}$ watts.

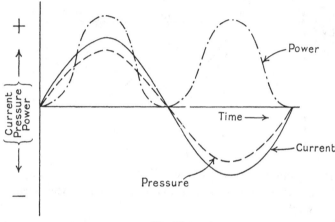

Fig. 39.

These are precisely the same formulae as would apply in a direct current case.

Power in a purely inductive circuit. Starting our consideration from the instant when current is zero (see Fig. 40), power will also be zero; then, as current increases positively, power will also increase positively attaining a maximum when the value of ie is a maximum; from this point the power commences to fall, attaining a zero value at the point c due to the pressure becoming zero. Going forward from the point c we find that current is still positive but pressure has become negative and therefore the sign of the power is also negative and we get a lobe of negative power from c to d; the shape of this negative portion of the curve is precisely the same, except as regards sign, as the previous positive portion. Passing on from the point d we now find that both current and pressure are negative with the result that power is again positive, the power variations during the second half cycle being precisely similar to those during the first half cycle.

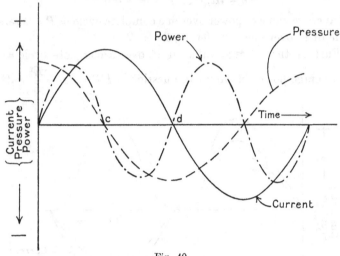

Fig. 40.

An examination of the curve will show that the average value of the power over a complete cycle is zero and the current is termed wattless or idle.

In most cases met with in practice we shall have a state of affairs intermediate in character between the two cases considered above; *i.e.* we shall have current differing in phase from the pressure by an angle lying between 0° and 90°. In such cases we can imagine the current split up into two components, one in phase with, and the other at right angles to, the pressure.

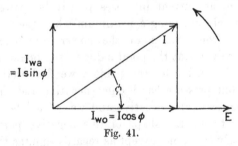

Fig. 41.

The total power of the current will be equal to the sum of the powers of its components; the magnitude of the component at right angles to the pressure is represented by the expression $I_{wa} = I \sin \phi$ and it evidently represents no power (see above, Fig. 40), it is termed the wattless component of the current; the component in phase with the pressure is $I_{wo} = I \cos \phi$ and the power represented by this is $IE \cos \phi$ which is also the power of the total current.

In order to obtain more complete and more exact ideas concerning power in alternating current circuits the matter can be treated mathematically.

If we have a current of I amperes driven by a pressure of E volts, the phase difference between them being $\phi°$, then

$$i = I_m \sin \theta, \quad e = E_m \sin (\theta + \phi)$$

and

$$p = ei = E_m I_m \sin \theta \, . \, \sin (\theta + \phi) = \frac{E_m I_m}{2} \{\cos \phi - \cos (2\theta + \phi)\}.$$

But $$\frac{E_m I_m}{2} = EI,$$

and therefore $p = EI \cos \phi - EI \cos (2\theta + \phi)$*.

We see then that the power at any instant is made up of two terms; one, $EI \cos \phi$, is fixed in magnitude during a cycle (*i.e.* it is not an alternating quantity) and depends upon the values of E, I and $\cos \phi$; the other, $EI \cos (2\theta + \phi)$, is an alternating quantity whose magnitude depends only upon E and I, and whose periodicity is double that of the pressure or current. If we take fixed values of E and I and look into the effect of altering the value of ϕ, we see that the former term alters in magnitude being greatest when E and I are in phase and zero when they differ in phase by 90°; the variation of ϕ does not alter the amplitude of the alternating term but alters its phase somewhat. The results may also be expressed by stating that for fixed values of E and I the shape of the power curve is invariable, but its axis lies above the axis of current and pressure by an amount depending upon ϕ. It is also noteworthy that the position of the maximum value of the power wave moves along as ϕ changes.

We can readily obtain the average value of the expression

$$p = EI \cos \phi - EI \cos (2\theta + \phi)$$

* $\cos \phi = \cos (-\phi)$.

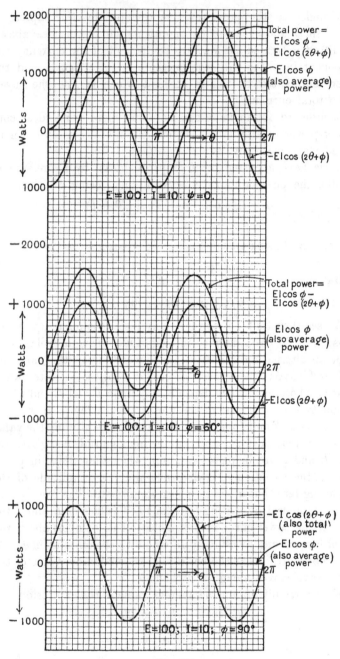

Fig. 42.

by noting that the first term is constant and therefore its average value is $EI \cos \phi$, while the average value of the latter term is obviously zero over a half cycle or any number of half cycles, hence, as before, we arrive at the expression $P = EI \cos \phi$ for the average power in a circuit in which E and I are the R.M.S. values of pressure and current and ϕ is the phase difference between them (it is quite immaterial whether current is leading or lagging).

The idea of a wattless current will at first sight appear very strange and it is perhaps desirable to examine, from a physical point of view, what is taking place in a circuit in which such a current is flowing. Imagine an alternator sending current to a choking coil whose resistance is negligible (other sources of loss also being neglected), the current sent will lag 90° behind the pressure and thus will be wattless. As the current rises (see Fig. 40) positively from its zero value, power is transmitted along the line to the choker and energy is being stored up in the magnetic field of the choker during this time. This goes on until the current attains a maximum value and the energy stored in the choker is also a maximum; then, as pressure reverses and the current begins to fall, the sign of the instantaneous power changes and this means that the choker is supplying power to the alternator and the energy stored in the magnetic field is gradually being given up again as the magnetic field diminishes in strength. When the current has fallen to zero the whole of the energy stored in the magnetic field of the choker has been given back to the alternator and the average power during the half cycle is zero. Thus we have energy oscillating between the alternator and the choker, the average power over any number of complete half cycles always being zero. Notice that in no case with single phase alternating current is power transmitted uniformly, it is always transmitted in a series of throbs which may result in flickering of lights and lack of uniformity of torque of motors.

Calculation of copper loss. The calculation of copper loss is a very important matter and for this purpose we can always make use of the formula $P = I^2 R$ whether the circuit is inductive or not. That this is so is due to the fact that the pressure component driving the current through the resistance

is always in phase with the current and therefore the copper
loss is equal to

Current × Pressure component due to resistance

= Current × Current × Resistance = I^2R.

Of course in applying this formula to windings it is necessary
to take care that the resistance is corrected for temperature and
for skin effect, if any. Both working and idle currents cause
copper loss in conductors and this is one of the reasons why idle
currents (due to low power factors) are avoided as far as possible
in practice. It is immaterial whether the copper loss is calculated
by using the total current or by using each component separately
and then adding, the following demonstration shows that the
same result is obtained in either case.

Loss due to idle component = $I^2{}_{wa}R = I^2R \sin^2 \phi$.

Loss due to working component = $I^2{}_{wo}R = I^2R \cos^2 \phi$.

Total loss = $I^2R (\sin^2 \phi + \cos^2 \phi) = I^2R$.

Power factor.

When the current and pressure in a circuit are not in phase
the product of those two quantities no longer gives the true
power in the circuit but gives what may be called the volt-
amperes of the circuit and the ratio

$$\frac{\text{True watts}}{\text{Volt-amperes}}$$

is termed the power factor of the circuit. In the case of sinusoidal
waves this ratio becomes $\dfrac{IE \cos \phi}{IE} = \cos \phi$, but it should be noted
that the power factor is only *strictly* equal to this expression in
the case of *sine* waves though it may be approximately true in
most practical cases.

Example. A pressure of 200 volts at 50 cycles per second is
applied to a circuit having an inductance of ·02 henry and a
resistance of 6 ohms, determine the power factor of the circuit and
the working and idle components of the current.

$$\text{P.F.} = \cos \phi = \frac{\text{Resistance}}{\text{Impedance}} = \frac{6}{\sqrt{(6·28 \times 50 \times ·02)^2 + 6^2}}$$

$$= \frac{6}{\sqrt{6·28^2 + 6^2}} = ·69.$$

$$\therefore \quad \phi = 46° \quad \text{and} \quad \sin \phi = ·719.$$

Total current $= \dfrac{200}{\sqrt{6 \cdot 28^2 + 6^2}} = 23$ amperes.

Working component $= I \cos \phi = 23 \times \cdot 69 = 15 \cdot 9$ amperes.

Idle component $= I \sin \phi = 23 \times \cdot 719 = 16 \cdot 5$ amperes.

Example. The power absorbed by a certain inductive circuit is 2000 watts when the current is 40 amperes, the applied pressure being 100 volts at 100 cycles per second. Determine the resistance and inductance of the circuit.

In this example the whole of the loss is copper loss and therefore we have

$$P = I^2 R \quad \text{or} \quad R = \frac{P}{I^2} = \frac{2000}{1600} = 1 \cdot 25 \text{ ohms.}$$

$$\text{Power factor} = \frac{\text{True watts}}{\text{Volt-amperes}} = \frac{2000}{4000} = \cdot 5 \, ;$$

$$\therefore \quad \phi = 60^\circ \quad \text{and} \quad \tan \phi = 1 \cdot 73.$$

Again $\tan \phi = \dfrac{\omega L}{R} = 1 \cdot 73 = \dfrac{628 L}{1 \cdot 25}$ whence $L = \dfrac{1 \cdot 73 \times 1 \cdot 25}{628}$

$$= \cdot 00344 \text{ henry.}$$

In connection with power factor a point occurs in relation to the rating of certain alternating current machines (alternators, transformers, etc.) which is perhaps worthy of special mention. In direct current work it is the usual custom to express the output of a generator in kilowatts (K.W.) but, in the cases mentioned, it is the almost invariable custom to express the output in kilovolt-amperes (K.V.A.), which can be found by dividing the kilowatts by the full load power factor on which the machine is intended to operate. This practice has arisen because the output of alternators and transformers is largely limited by questions of heating*, and the heating of these machines is more dependent upon the K.V.A. (which is proportional to the total current through the machine) than upon the K.W. (which is proportional to the working component).

Thus, if we purchase a transformer rated at 10 K.V.A. we can

---

* In connection with the permissible output of a machine the question of pressure regulation is also of great importance but we are not at present in a position to consider this matter. It is worthy of mention however that from this point of view also the permissible output of a machine is better expressed by stating the K.V.A. than by stating the K W.

rest assured that we shall be able to obtain 10 K.V.A. out of it
(so far as heating is concerned) no matter on what P.F. the machine
may be worked; but if we invested in one rated at 10 K.W. and
obtained this power with the maximum permissible temperature
rise at unity P.F., the temperature rise would be far too high if
we attempted to take 10 K.W. from the same machine at ·8 P.F.
As a matter of fact in this case we should be trying to take 12·5
K.V.A. from the machine and should obtain a corresponding temperature rise. The same reasoning will also apply to the armature
circuit of an alternator and to some extent, though in an indirect
manner involving questions of regulation, to the field circuit.

Table of power factors met with in practice.

Lighting systems	·8	to 1·0
Power systems	·5	to ·9
Artificial condensers	·004	to ·01
Mercury rectifiers (fully loaded) ..	·7	to ·9
Induction motors (running light) ..	·1	to ·2
Induction motors (loaded)	·8	to ·9

*Measurement of power and power factor in alternating current
circuits.*

These measurements may be effected in several ways and
first the three ammeter and three voltmeter methods should
be noted. These methods are dealt with on account of the fine
examples they afford of the use of vector diagrams rather than
because of their practical importance at the present time.

The three ammeter method. In order to make use of this method
of determining the power in an inductive circuit a non-inductive
resistance should be placed in parallel with the circuit in question,
three ammeters being arranged to measure the current in each
of the paths and also the total current (as indicated in Fig. 43).
The total current will of course be less than the arithmetical sum
of the two components on account of the phase difference existing
between the latter. The vector diagram for the circuit will be
as indicated in the figure, the pressure applied to the circuit
being used as the vector of reference (that is the one drawn
first and from which the others are measured).

Now power absorbed in inductive circuit

$$= I_1 \times E \times \text{p.f.} = I_1 \times E \times \cos\phi = I_1 \times E \times \cos GOC.$$

Also $\cos GOC = \cos BCD = -\cos OCD = -\left(\dfrac{I_1{}^2 + I_2{}^2 - I_3{}^2}{2I_1 I_2}\right)$ *

$$= \frac{I_3{}^2 - I_1{}^2 - I_2{}^2}{2I_1 I_2}.$$

∴ Power absorbed in inductive circuit

$$= I_1 \times E \times \frac{I_3{}^2 - I_1{}^2 - I_2{}^2}{2I_1 I_2} = \frac{E}{2I_2}(I_3{}^2 - I_2{}^2 - I_1{}^2) \text{ watts.}$$

If the value of the non-inductive resistance be known then $R = \dfrac{E}{I_2}$ and power in inductive circuit $= \dfrac{R}{2}(I_3{}^2 - I_2{}^2 - I_1{}^2)$ watts.

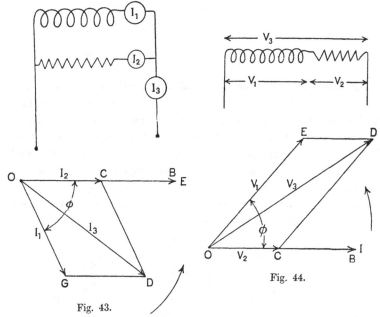

Fig. 43.

Fig. 44.

The three voltmeter method. In this case a non-inductive resistance is inserted in series with the circuit in which the power is to be measured, the pressure being taken across each part and the complete circuit by three voltmeters or, which is perhaps better, by a single voltmeter with the help of a three way voltmeter switch. The vector diagram for this case is shown in Fig. 44

* See chapter on solution of triangles in text book of trigonometry.

and is best constructed by taking the current as the vector of
reference.

Now power absorbed in inductive circuit under test

$$= V_1 \times I \times \text{P.F.} = V_1 \times I \times \cos BOE.$$

Also $$\cos BOE = \cos BCD = -\cos OCD$$

$$= -\left(\frac{V_2{}^2 + V_1{}^2 - V_3{}^2}{2V_2 V_1}\right) = \frac{V_3{}^2 - V_1{}^2 - V_2{}^2}{2V_2 V_1}.$$

$\therefore$ Power absorbed $$= V_1 \times I \times \frac{V_3{}^2 - V_1{}^2 - V_2{}^2}{2V_2 V_1}$$

$$= \frac{I}{2V_2} \times (V_3{}^2 - V_1{}^2 - V_2{}^2) \text{ watts},$$

or, if the resistance of the non-inductive circuit is known,

$$R = \frac{V_2}{I} \quad \text{or} \quad \frac{I}{V_2} = \frac{1}{R};$$

$$\therefore \text{ Power} = \frac{1}{2R} \times (V_3{}^2 - V_1{}^2 - V_2{}^2) \text{ watts}.$$

Wattmeter method for measuring power and power factor. This
method is simple and, if a good instrument is available, accurate,
and is by far the most commonly used method for the purpose
in question. The theory and construction of wattmeters is given
in Chap. VI, and at present it is only necessary to note that such
an instrument possesses two coils; through one, known as the
current coil, the main current, or a definite proportion of the
main current, or a current proportional to the main current,
passes; and through the other, known as the pressure coil, a

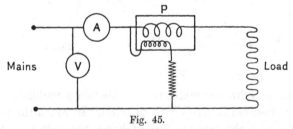

Fig. 45.

current passes which is proportional to the pressure across the
circuit. These coils should be connected relatively to each other
as shown in Fig. 45 which is drawn on the assumption that the
whole of the main current passes through the current coil, and

that a current proportional to the pressure passes through the pressure coil owing to the fact that it is connected across the mains in series with a non-inductive resistance. The power is read directly on the meter and the power factor may be calculated with the help of the relationship

$$\text{Power factor} = \frac{\text{Watts}}{\text{Amperes} \times \text{Volts}} = \frac{\text{Watts}}{\text{Volt-amperes}}.$$

Power factor meter method for measuring power and power factor. Occasionally, on switch boards, a power factor meter (*i.e.* an instrument for directly indicating the power factor of the circuit) is fitted rather than a wattmeter, and in this case the power passing through the circuit may be obtained by using the formula

Power = Current × Pressure × Power factor.

The power factor meter is, for single phase working, provided with two pressure coils and a current coil, and the external connections are precisely the same as those shown for the wattmeter in Fig. 45.

Use of condensers and choking coils for adjusting the pressure applied to pieces of apparatus. The wattless properties of currents, lagging or leading in regard to the pressure by a quarter of a period, may be taken advantage of in order to cut down the pressure applied to an appliance without wasting an undue amount of energy. Thus suppose we wish to run a single arc lamp, needing say 80 volts, off a 110 volts circuit; if the circuit supplied direct current we should have to absorb the surplus pressure by placing a suitable resistance in series with the lamp and this would involve considerable waste of energy, but, on an alternating current circuit, a choking coil (*i.e.* a coil designed to have a considerable reactance and the minimum resistance) could be inserted in series with the lamp and this would absorb the superfluous pressure just as effectively as the resistance but would absorb very little power*.

* Some little power would, in practice, be absorbed in the choker owing to copper and core losses but the amount would be small, in other words the power factor would be very low but not actually zero. Though the case of the arc lamp has been cited as one in which a choker might with advantage be employed, an auto-transformer would most likely be used in practice since this would not only prevent waste of power but would also diminish the current taken from the mains and give a better power factor.

When a choker is used to cut down the pressure applied to a non-inductive circuit it is clear that the sum of the squares of the pressures across the choker and N.I. circuit is equal to the square of the line pressure.

Choking coils are also very useful in experimental and testing work as they possess, in addition to the advantage already mentioned, the property of giving a perfectly continuous variation of current (as distinct from the step by step variations produced by the ordinary resistance) if provided with a movable core.   The author has made considerable use, for experimental purposes, of chokers composed of 440 turns of number 11 D.C.C. copper wire wrapped on a wooden former and provided with tappings at intermediate points in order to increase the range.   The core consists of iron stampings, $12'' \times 3''$, insulated by a layer of

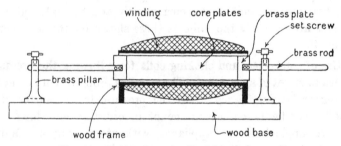

Fig. 46. Section through adjustable choking coil.

varnish on one side and secured, by insulated bolts, to a central brass plate to which are riveted two brass rods which pass through the pillars at either end, adjustment of the choking effect by altering the core position being thus arranged for (see Fig. 46). Each pillar is provided with a set screw in order that the core may be clamped in any desired position, a precaution which is very necessary when large currents are being used owing to the tendency of the core to take up a position in which there is a maximum number of magnetic lines passing through it (*i.e.* a central position).

At first it might be thought that the comparatively massive brass centre piece would be the seat of considerable eddy currents, which would cause considerable loss, but this has not been found to be the case in practice because the greater part of the flux passes through the iron core and the flux density in the brass is very low. The construction indicated has been found excellent from the mechanical point of view and the power factor obtained is of the order of ·03 to ·15 depending on the core position. Such a choker is capable of absorbing 5 K.V.A. continuously and 10 K.V.A. for a short time; the capacity can of course be increased by increasing the section of the copper wire used or by increasing the section of the core.

Condensers can also be used to cut down the pressure without wasting much power; Mr Ashton has suggested methods, involving the use of condensers, whereby low voltage and low candle power lamps may be run off high pressure (250 volts) circuits, thus obtaining the advantages of using stronger lamps and, at the same time, lower candle powers, than would otherwise be possible*.

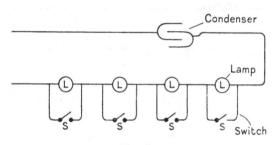

Fig. 47.

Perhaps the most interesting system is that in which each circuit consists of a number of lamps in series with each other and with a condenser, each lamp being fitted with a short-circuiting switch which is normally closed and which may be opened when it is desired to switch on the lamp. All the lamps used in any one series circuit should be arranged to take the same current, any variation in candle power in different lamps being obtained by using filaments requiring more or less pressure for higher or lower candle powers respectively. The total pressure taken by the maximum number of lamps in at any one time should not exceed 40 % of the supply pressure, and the condenser should be of such a size that when lamps whose aggregate pressure amounts to 25 % of the line pressure are switched in they obtain the correct current

* *J. I. E. E.*, Vol. 49, p. 703

and also, of course, the correct pressure. If fewer lamps are in
they will be slightly over run, and if more lamps are in they
will be under run, the departure from the correct current is not,
however, very great.

This system, in addition to the advantages already mentioned,
causes a leading current, which is excellent from the central
station point of view, and prevents an excessive rush of current
occurring when the cold metallic filament is first switched on to
the mains.

The British Insulated and Helsby Cable Cos., Ltd., (to whom
the author is indebted for the following information and for
Figs. 48 and 49) have also developed the use of Mansbridge
condensers to cases where it is desired to supply a few metallic
lamps off alternating current circuits of medium pressure. In

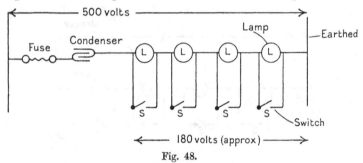

Fig. 48.

such cases it is of the utmost importance to prevent an unduly
high pressure developing across the lamps or between the lamps
or switches and earth.   Thus, for a pressure of 500 volts, the scheme
of connections shown in Fig. 48 may be used; an examination of
the circuits will show that if a lamp filament breaks the full
pressure of 500 volts will occur across the break so long as the
switch remains open, the normal distribution of pressure re-
asserting itself as soon as the switch is closed.

If greater safety is desired the arrangement of Fig. 49 may be
used; in this case the removal of a lamp or a break in the filament
does not cause an interruption of the circuit since there are two
condensers in series right across the mains. If the condenser
B is of considerably higher capacity then the condenser A a break
in the lamp circuit will not cause the development of a seriously

high pressure across the lampholder or between the switches and earth. An automatic earthing device can also be fitted which will prevent danger should the condenser *A* become short-circuited in any way.

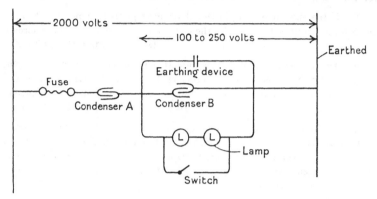

Fig. 49.

EXAMPLES

1. If a pressure of 200 volts at 50 cycles per second is applied to a circuit whose inductance is ·03 henry and resistance 12 ohms, calculate the current flowing and the power wasted in copper loss.

Answer. 13·1 amperes; 2060 watts.

2. Calculate the working and idle components of the current and the power factor for the circuit described in the above example.

Answer. 10·3 amps.; 8·06 amps.; ·788.

3. When a pressure of 100 volts is applied to an inductive circuit the current flowing is 10 amperes and the power absorbed 800 watts. Calculate the resistance and reactance of the circuit.

Answer. 8 ohms; 6·00 app. ohms.

4. If 100 volts at a frequency of 100 cycles per second is applied to the circuit mentioned in example (1), calculate the K.W. and K.V.A. absorbed.

Answer. ·24 K.W.; ·447 K.V.A.

5. A non-inductive resistance taking 24 amperes is connected in parallel with an inductive circuit taking 16 amperes. If the resultant current is 33 amperes, determine the power factor of the inductive circuit.

Answer. ·334.

6. If the pressure applied to the circuit in the above example is 110 volts determine the power absorbed in the inductive portion. *Answer.* 588 watts.

7. In testing the power factor of a circuit by the three voltmeter method the total applied pressure was 220 volts, the component across the non-inductive resistance being 140 volts and that across the inductive resistance being 150 volts. Determine the phase angle between current and pressure for the inductive and for the complete circuits.   *Answer.* 81°; 42°.

8. If the current flowing in the above example is 8 amperes, determine the total power absorbed and also the power absorbed in the inductive circuit.
Answer. 1300 watts; 180 watts.

9. The power absorbed in an inductive resistance is 4000 watts, the current being 48 amperes and the pressure 220 volts. Determine the power factor of the circuit and the working and wattless components of the current.
Answer. ·379; 18·2 amps.; 44·5 amps.

10. A number of 20 volt, 10 watt lamps are to be connected in series with each other and with a condenser and run off a 200 volt circuit at 100 cycles per second. Determine the capacity of the condenser necessary in order that when two lamps are switched in they may each be supplied with their correct pressure and current.   *Answer.* 4·06 m.f.

11. If, in the above case, only one lamp is switched in, determine the pressure across it. (Assume the resistance of the lamp to remain constant.)
Answer. 20·3 volts.

12. Repeat example (11) for the case when three lamps are switched in circuit.   *Answer.* 19·5 volts per lamp.

CHAPTER V

MULTIPHASE CURRENTS

Multiphase currents are perhaps best approached from the point of view of the generator: consider the case of a ring (as a matter of fact the remarks also apply to a drum winding) armature situated in a two pole field and having a uniformly distributed winding, let this be provided with two slip rings connected to opposite points of the winding. It is clear that there will be two parallel paths through the armature and at each and every instant the pressures produced in these two paths will exactly balance each other as regards the internal circuit of the armature. Considering one of the paths only, it will be evident that to get the maximum resultant pressure from the conductors in series, it is necessary that all the small component pressures in the individual conductors should be in phase with each other and this cannot possibly be the case with a winding distributed to the extent indicated above. If we concentrate the conductors composing one path, so that they occupy but a small portion of the armature core, we shall bring their pressures more nearly into phase with each other, but this procedure will again diminish the output of the machine since we are no longer making use of the whole of the periphery of the armature and hence must reduce either the number or the cross-section of the conductors. In practice a compromise is arrived at and in single phase machines we usually find that about five-eighths of the armature periphery is used for winding purposes. This remark applies to windings which are solely used for the production (or utilisation) of alternating currents; in the case of windings dealing with both alternating and direct currents, as in rotary converters, a uniformly distributed winding must be used on account of considerations arising from the direct current side.

It is perhaps desirable to point out that windings used solely for alternating currents differ from those used for direct currents in one other important respect. In order to get sparkless commutation direct current windings are invariably of the closed circuit type in modern machines, that is, there is a closed internal circuit through the armature winding which means there will be at least two paths from brush to brush; in alternating current windings this arrangement is no longer necessary and it is therefore usual, except perhaps in very heavy current machines, to have the whole of conductors of one phase connected in series. It may be pointed out that when a concentrated winding is used for a single

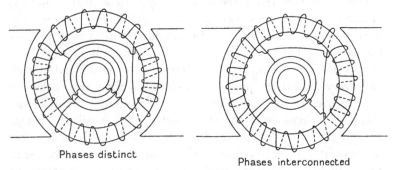

Phases distinct

Phases interconnected

Fig. 50.

phase machine the unoccupied portion of the armature is not completely wasted, it is available for the dissipation of heat. In order to make use of the spare space on the armature, and at the same time avoid (to any serious extent at any instant) the production of opposing pressures in conductors connected in series, we may place on the armature two windings which, for the moment, may be regarded as quite distinct, that is, they are insulated from each other, provided with separate slip rings and used to feed current into two distinct circuits (see Fig. 50). A little consideration will show that one of these currents will be at its maximum value when the other is passing through its zero value, that is, they differ in phase by a quarter of a period or 90°; we have, in fact, what are usually known as two phase currents. It is not absolutely essential that the two phases be kept quite distinct, in many cases it may be permissible to join one end of one phase to one end of the other and, since this may be effected inside the machine, but three slip rings would then be required and three line wires would be used.

Let us investigate the magnitudes and relative phases of the several quantities concerned in a two phase inter-connected system and for this purpose assume that the phases are balanced, that is, the pressure, current and phase difference between current and pressure in one phase are equal to the corresponding quantities in the other phase. The vector diagram for such a case is shown in Fig. 51, where E_1 and E_2 represent the two phase pressures, I_1 and I_2 the two phase currents and ϕ the phase difference between current and pressure in each phase.

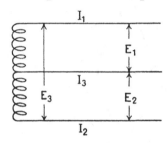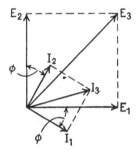

Fig. 51.

The pressure between the two outer wires (E_3) is compounded of two phase pressures which are at 90° apart and thus is readily seen to be $\sqrt{2}$ or 1·414 times the pressure produced in one phase. Similarly the current in the common wire (the common wire should not be referred to as the middle wire as this would be likely to cause confusion with the three wire D.C. method of distribution, from which the case under consideration is quite distinct) is compounded of two phase currents, which are also separated by a phase difference of 90°, and hence is equal to 1·414 times the current due to one phase. If the phases are not balanced the above relationships will not hold but the actual magnitudes and phases of the currents and pressures can readily be worked out from a vector diagram in any case for which the necessary information is given.

Instead of being content with two windings on the armature we may employ three, each occupying one-third of the total periphery of the armature, and with the breadth of each of the two bunches of conductors, into which each winding is divided, equal to one-third of the pole pitch (see Fig. 52).

We shall now produce what are known as three phase pressures and currents and, if desired, the three circuits can be kept quite distinct thus necessitating the use of six slip rings; in practice, however, the three phases are almost invariably inter-connected in one of the ways mentioned below.

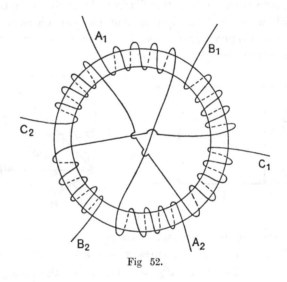

Fig 52.

In order to grasp clearly the principles underlying the inter-connection of the phases in a three phase system, it is necessary to have sound ideas concerning the relative phases and magnitudes of the pressures and currents produced, and these will of course depend on the precise way in which the phases are connected. In Fig. 52 consider the inter-connection of phases A and B; if we pass through phase A from A_1 to A_2 and then join on to B, passing through this from B_1 to B_2, the two pressures will be at 60° apart and the resultant pressure will be shown at D in Fig. 53. If, however, after passing through A we connect on to B and pass through this from B_2 to B_1 the two pressures will be at 120° apart and the resultant will be shown at E, we have, in fact, in the latter case reversed B relative to A as compared with the former case. It is evident therefore that the phase and magnitude relations in inter-connected three phase systems depend very

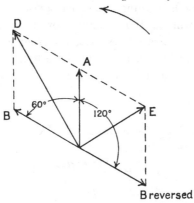

Fig. 53.

largely on the precise ends of the several phases which are connected together and the utmost care is necessary in this respect. The scheme of inter-connections will probably be clearer if we consider a slightly modified winding arrangement as shown in Fig. 54, where the two groups of conductors forming one phase

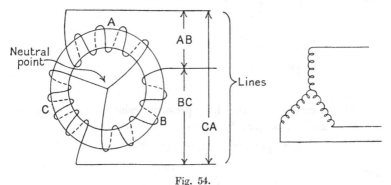

Fig. 54.

are grouped together; this would, in practice, result in rather considerable phase differences in the pressures produced in conductors connected in series with a resultant lowering of the total pressure produced per phase.

Star method of inter-connection of three phase windings. This is accomplished by connecting the front ends of each winding together to form what is known as the neutral point, the three remaining ends being connected to the three line wires (see Fig. 54). If we consider each winding taken clockwise in this figure the three pressures will be 120° apart; now each line pressure is compounded of two phase pressures, the front end of one phase being connected to the back end of the other, and therefore the two pressures must actually be added at 60° in order to obtain the line pressure.

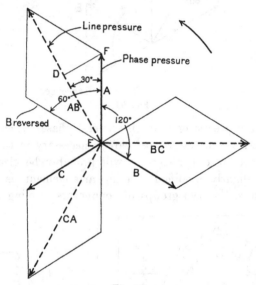

Fig. 55.

The magnitude of the line pressure relative to the phase pressure can now be readily found as indicated below.

$$\text{Line pressure} = 2DE \quad (\text{see Fig. 55}),$$

but
$$\frac{DE}{FE} = \cos 30°,$$

$$\therefore \quad \text{Line pressure} = 2FE \cos 30°$$

$$= 2 \times \cos 30° \times \text{Phase pressure} = 2 \times \frac{\sqrt{3}}{2} \times \text{Phase pressure}$$

$$= \sqrt{3} \ \text{Phase pressure} = 1{\cdot}73 \times \text{Phase pressure}.$$

The line current is obviously equal in magnitude and phase to the corresponding phase current.

It will be noticed in Fig. 54 that no connection is shown to the neutral point, this is not necessary so long as the load on the system is truly balanced since the total current flowing through the three lines can at each instant be shown to be zero and thus, even if a connection was provided between the neutral point of the load and the neutral point of the generator, no current would flow through it*.

Resultant current in neutral conductor (if used)

$$= I_m \sin\theta + I_m \sin(\theta + 120) + I_m \sin(\theta - 120)$$
$$= I_m (\sin\theta + \sin\theta\cos 120 + \cos\theta\sin 120 + \sin\theta\cos 120 - \cos\theta\sin 120)$$
$$= I_m (\sin\theta + 2\sin\theta\cos 120) = I_m (\sin\theta - \sin\theta) = 0.$$

If the load is connected in star and the phases are unbalanced, as may easily be the case in lighting systems, a neutral conductor must be provided though its section may be less than that of either of the other three line wires.

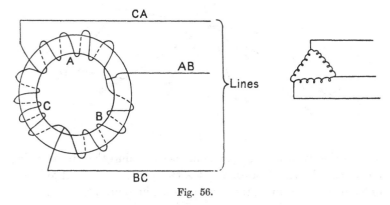

Fig. 56.

Mesh or Delta inter-connection of three phase circuits. Considering the three phase winding arrangement in which, for the two pole case, the whole of the conductors of one phase are in one band (not a desirable arrangement for the reason already given), let the three phases be connected internally as in Fig. 56, the line wires being taken off at the junctions of the phases as

* This statement is made on the assumption that the pressure waves are truly sinusoidal.

shown. In the first place we notice that there will be no resultant pressure (at any rate in the case of sinusoidal waves) round the internal mesh circuit; this may be demonstrated in a similar manner to that used to show the absence of current in the neutral conductor of a three phase balanced system connected in star. Next we notice that the pressure between lines (*i.e.* the line pressure) is equal to the phase pressure. As regards the relationship between the line and phase currents, we see that each line current is compounded of two phase currents and, since current is taken off from the back end of one phase and from the front end

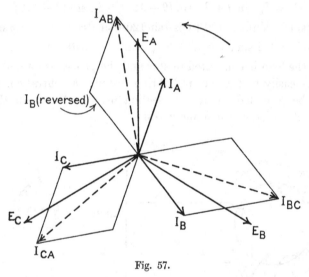

Fig. 57.

of the other (which we have already seen is equivalent to reversing one of the phases), we must add the two phase currents at an angle of 60° in order to obtain the corresponding line current. Proceeding in a manner similar to that used in obtaining the relationship between line and phase pressures in the case of star connection, we find that the line current is $\sqrt{3}$ or 1·73 times the phase current in the case of a balanced mesh connected system.

A vector diagram for a three phase balanced mesh connected system is shown in Fig. 57.

Both star and mesh methods of inter-connection are extensively employed in practice, the star system of course being especially suitable when high line pressures are required. For alternators, induction motors and the primaries of step down transformers, the star system is perhaps best adapted and is most commonly employed, while mesh connection is frequently used on the low tension side of lighting transformers and of course in the windings of machines intended to deal with both alternating and direct currents, as in the case of rotary converters.

There is one trouble that is sometimes met with in mesh connection that is worthy of special mention and that is the risk of unbalanced pressure round the internal circuit of the mesh. The fundamental sine waves will always neutralise each other round the internal circuit but, in practice, harmonics may be present and certain of these may, in connection with the internal circuit, be cumulative and thus give rise to higher frequency circulating currents which will cause undue heating. This will happen in the case of a system which has the third harmonic present in each wave, the phase relationship between the harmonic and fundamental being of course the same for each phase. The matter is illustrated by a wave diagram in Fig. 58, the neutralisation of the fundamental and the cumulation of the third harmonic, giving rise to a triple frequency circulating current round the mesh, being clearly seen.

The same general effect is noticeable in the case of certain other harmonics as, for instance, the ninth and fifteenth, and, in a three phase machine which came under the notice of the writer, the presence of the ninth harmonic gave rise to a resultant pressure round the mesh of about 30 volts, the line pressure of the machine being 170 volts.

In conclusion it is perhaps desirable to point out that combinations of star and mesh can, if necessary, be used, but these are of comparatively rare occurrence.

The relative merits of single and multiphase systems depend upon the purpose to which the power is to be applied: in the following comparison three phase current is taken as typical of multiphase systems since this is the case most commonly met with in practice.

Single as against multiphase currents for motors. If the motor is required to have a series characteristic, as for traction or crane work, single phase current, used in connection with a commutator motor, is quite satisfactory; if a shunt characteristic is required (*i.e.* approximately constant speed at all loads) then induction motors are best and the three phase induction motor is a much more satisfactory machine than the single phase motor, being cheaper for the same output and having much better starting properties.

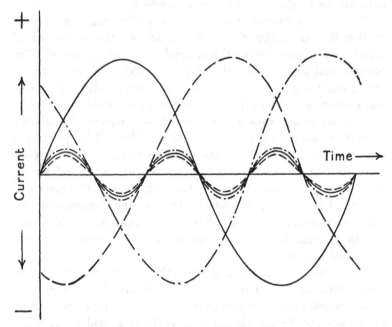

Fig. 58.   For the sake of clearness the amplitudes of the harmonics are shown slightly different, in practice they would be equal.

Single as against multiphase currents for lighting purposes. In this connection single phase lamps are almost invariably employed even on multiphase systems and though a three phase arc lamp has been evolved, which would seem to remedy the flickering so noticeable with single phase lamps, yet it has not been extensively employed and we may conclude that so far as this application is concerned there is nothing to choose between the two systems.

Single as against multiphase currents for transmission purposes. For this application a careful comparison will show that there is considerable advantage to be gained by using multiphase currents,

the advantage usually taking the form of a diminution in the
amount of copper used. The exact amount of copper which can
be shown to be saved depends upon the basis of comparison, thus,
if we make our comparison on the basis of equal losses in the two
systems, we shall be able to show a greater saving than if we make
our comparison on the basis of equal current densities in the
conductors used ; in practice the actual conditions would sometimes
be best met by one form of comparison and sometimes by the
other and both cases are included in the following table, which
shows the relative amounts of copper required in the different
systems when transmitting 1000 k.w. at a line pressure of 10,000
volts, unity power factor being assumed in each case.

Type of transmission	Number of lines	Current per phase (amperes)	Current per line (amperes)	Cross-section per line (sq. in.)	Total cross-section of conductor (sq. in.)	Relative total cross-section
Single phase	Two	100	100	·1	·2	1·00
Three phase	Three	33·3	57·7	·0577	·1731·	·865*
Three phase	Three	33·3	57·7	·05	·15	·75†

In the above example mesh connection has been assumed in
each of the three phase cases for the sake of convenience but, as
a matter of fact, the mode of connection is quite immaterial so
long as the line pressure is the same in each case. The working
out for the case of equal current densities is quite simple and needs
no comment and the section of the conductor for the case of equal
losses is found as follows :

Let I and R be the current and resistance per line in the single
phase case, and I_1 and R_1 be the corresponding values in the three
phase case then, for equal total losses, we have

$$2I^2R = 3I_1^2R_1 \text{ or } R_1 = \frac{2I^2R}{3I_1^2} .$$

For the moment assume the resistance R to be unity, then

$$R_1 = \frac{2 \times 100^2 \times 1}{3 \times 57\cdot8^2} = 2.$$

Now if the resistance of a single phase line is one-half that
of a three phase line it follows that the section of the latter is half

* Equal current densities in all conductors, namely 1000 amperes per sq. in.
† Equal losses in conductors and equal efficiencies of transmission.

that of the former and is therefore ·05 sq. inch in the case under consideration.

Measurement of power and energy in multiphase circuits. In practice the only methods that need be considered are those involving the use of a wattmeter in the case of power, and a supply meter in the case of energy, and our chief point of interest at present is in the correct connection of these instruments to the circuits.

It is desirable to realise at once that any connection suitable for the measurement of power is also suitable for the measurement of energy if a supply meter is substituted for the wattmeter used in the former case.

The type of supply meter which is almost invariably used is a true energy meter (as distinct from a coulomb meter), this is necessary because the variation of power factor causes the energy to be no longer even approximately proportional to the product of current and time, and, since it is therefore necessary to introduce a pressure coil in order that the differences in phase between pressure and current may be taken into account, it is obviously a very simple matter to arrange that the variations of pressure also be allowed for.

It is clear that if ordinary energy meters are used the wattless component of the current is not charged for at all and this is perhaps hardly fair, since wattless currents certainly involve additional outlay in generators and mains, and also cause increased copper losses in transmission : recently attention has been directed to this matter and the idea of charging consumers for wattless current, in addition to the energy consumed, has been seriously considered*.

Connections of meters for the measurement of power and energy in two phase circuits. When the load is balanced an ordinary single phase meter can be inserted in one phase and the reading doubled, either by specially figuring the scale or otherwise (see Fig. 59 (*a*)). If the phases are not balanced it will be necessary to use two instruments, one being placed so as to indicate the power (or energy) of the first phase and the other the corresponding quantity of the second phase, the final result then being obtained

* See *J. I. E. E.*, Vol. 51, p. 270.

by the addition of the two readings.   If desired the two movements
may be placed on one spindle and the scale calibrated so as to
read the total power (or energy) in the circuit; in such an arrangement great care must be taken to secure that the accuracy of one
movement is not affected by the proximity of the other (see
Fig. 59 (*b*)).

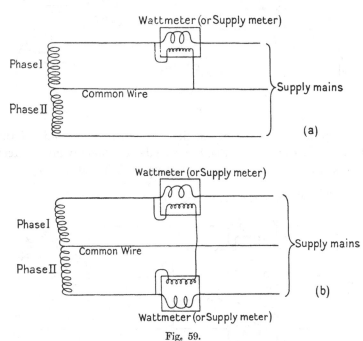

Fig. 59.

Measurement of power and energy in three phase circuits.   In
three phase circuits the above quantities are readily measured if
the circuit is balanced and the neutral point accessible; in this
case one wattmeter may be used, having its current coil in one
line and its pressure coil connected between that line and the
neutral point, as indicated in Fig. 60 (*a*), the total power (or energy)
being obtained by multiplying the instrument reading by three.
If the neutral point is not available an artificial one may be
created by connecting one end of each of three equal resistances,
one of which may be the pressure circuit of the wattmeter, to
each line, the remaining ends of the three resistances being starred

as in Fig. 60 (*b*). If the load is unbalanced three wattmeters may
be used, each having its current coil in one line and its pressure
coil between that line and the neutral point.   In this case if the
neutral point is not available an artificial one may be created, if

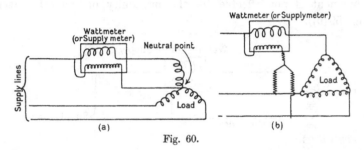

Fig. 60.

the wattmeters are exactly similar, by starring the ends of the
three pressure coils as in Fig. 61; this device will only give correct
results if the resistances of the pressure coils are all equal.

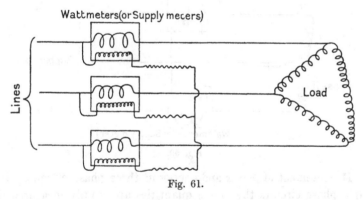

Fig. 61.

In approaching the most usual method used for the measurement of power (and energy) in three phase circuits, it is perhaps
desirable to deal with the matter in a rather more general way, since
the method is essentially suitable for any number of phases as
three, four, six or twelve.   If there are n phases $(n-1)$ wattmeters
are necessary each with its current coil in one line and its pressure
coil between that line and the line which contains no current coil.
Fig. 62 shows the connections for a four phase star connected
system in which case three wattmeters will be necessary.

At any instant let the currents and pressures in each of the phases be represented by i_1, e_1; i_2, e_2; i_3, e_3 and i_4, e_4 respectively, then the currents in, and the pressures between, the lines will be as indicated in the figure.

(In obtaining the values of the pressures between the lines a phase pressure is taken as positive when a coil is passed through towards the neutral point and *vice versâ*.)

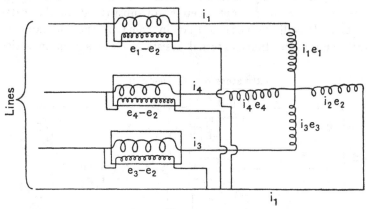

Fig. 62.

The total power at the instant under consideration will be
$$i_1 e_1 + i_2 e_2 + i_3 e_3 + i_4 e_4,$$
and the power indicated by the wattmeters will be
$$i_1 (e_1 - e_2) + i_4 (e_4 - e_2) + i_3 (e_3 - e_2)$$
$$= i_1 e_1 + i_4 e_4 + i_3 e_3 - e_2 (i_1 + i_2 + i_3).$$
But at every instant
$$i_1 + i_2 + i_3 + i_4 = 0 \quad \text{or} \quad i_2 = - (i_1 + i_2 + i_3),$$
and therefore the total power indicated by the wattmeters will be
$$i_1 e_1 + i_2 e_2 + i_3 e_3 + i_4 e_4,$$
which is also the total power in the four phases at the same instant. Since the above investigation is true at every instant the sum of the average readings of the wattmeters will be equal to the average total power in the phases. A similar method of proof can also be used if the phases are in mesh. Adapting the general proposition to the case of three phase circuits, whether

in star or mesh, we see that the total power can be measured, whether the circuits are balanced or not, by two wattmeters connected as in Fig. 63. Instead of using two distinct instruments one may be used having the two movements acting on a single spindle and pointer but, as in the two phase case, care must be taken that inaccuracies do not result from the interaction of one movement on the other.

The method may obviously be used to measure the energy taken by a circuit if supply meters are substituted for the wattmeters. It will be noticed that in some of the diagrams of wattmeter connections external resistances are shown in series

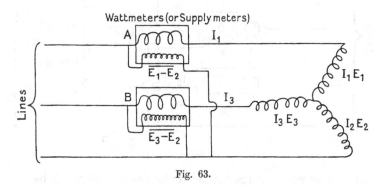

Fig. 63.

with the pressure coils of the instruments and in other cases none are shown, it is not intended that any difference of construction or operation should be indicated by this distinction and, as a matter of fact, both types are met with in commercial instruments, the necessary series resistance sometimes being external to, and sometimes contained within, the instrument.

The above method of connection is very useful and very commonly employed, both for wattmeters and for supply meters, and will repay a more detailed examination. Let us investigate the behaviour of two wattmeters, connected as above, when the power factor of the circuit is gradually lowered from unity to zero, the circuit being supposed to be balanced or at any rate approximately so. An appropriate vector diagram is shown in Fig. 64 where E_1, E_2 and E_3 represent the phase pressures, $(E_1 - E_2)$ and $(E_3 - E_2)$ then represent the pressures on the pressure coils of the two wattmeters A and B respectively. At first let I_1, I_2 and I_3, representing the phase currents (and also the line currents for star connection), be in phase with their respective phase pressures, then, since the angle between I_1 and $(E_1 - E_2)$ is equal to the angle between I_3 and $(E_3 - E_2)$, the two wattmeters will read alike. As the currents lag more and more behind their respective pressures θ_1 increases and θ_2 decreases, thus, if the total power is supposed to remain constant, the reading on A falls while that on B rises. When each phase current lags 60° behind

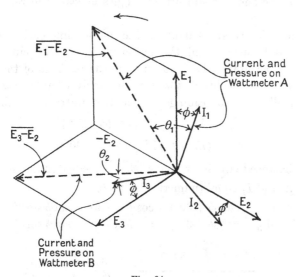

Fig. 64.

the corresponding phase pressure θ_1 will be 90° and therefore the wattmeter A will read zero, the reading on the other wattmeter continuing to rise if the total power is still supposed to remain constant. Since cos 60° is equal to ·5, wattmeter A will read zero when the power factor of the circuit is ·5, current lagging. Still greater angles of lag will cause wattmeter A to read backwards and its reading must then be subtracted from the other in order to obtain the true power. Such a state of affairs is likely to happen when the circuit consists of induction motors or transformers on little or no load, and great care must be taken that the wattmeter readings be not added when they should be subtracted.

A good plan is to notice the readings of the instruments when the induction motor is fairly well loaded, the power factor is then practically certain to be greater than ·5, and both wattmeters should read forward and if, on a diminished load, one instrument commences to read backwards, its reading should be subtracted from the reading of the other instrument in order to obtain the true power.

A further study of the vector diagram will show that if the phase current leads the phase pressure by 60° the wattmeter B will read zero and, for still greater angles of lead, will read backwards.

If the phases are balanced, the phase difference between pressure and current and the power factor of the circuit can readily and accurately be deduced from the readings of the wattmeters with the help of the formula arrived at below. Let W_A and W_B be the readings of the two wattmeters, then

$$W_A = (\overline{E_1 - E_2}) \cdot I_1 \cdot \cos(\phi + 30°),$$

and $\qquad W_B = (\overline{E_3 - E_2}) \cdot I_3 \cdot \cos(\phi - 30°).$

For balanced circuits we have $(\overline{E_1 - E_2})$ equal (numerically) to $(\overline{E_3 - E_2})$, and I_1 equal (numerically) to I_3.

$$\therefore \frac{W_A}{W_B} = \frac{\cos(\phi + 30°)}{\cos(\phi - 30°)} = \frac{\cos\phi \cos 30° - \sin\phi \sin 30°}{\cos\phi \cos 30° + \sin\phi \sin 30°}$$

$$= \frac{\cos\phi \times \dfrac{\sqrt{3}}{2} - \sin\phi \times \frac{1}{2}}{\cos\phi \times \dfrac{\sqrt{3}}{2} + \sin\phi \times \frac{1}{2}} = \frac{\sqrt{3}\cos\phi - \sin\phi}{\sqrt{3}\cos\phi + \sin\phi};$$

$$\therefore W_A \sqrt{3}\cos\phi + W_A \sin\phi = W_B \sqrt{3}\cos\phi - W_B \sin\phi,$$

or $\qquad \cos\phi(\sqrt{3}\,W_B - \sqrt{3}\,W_A) = \sin\phi(W_B + W_A),$

and $\qquad \dfrac{\sin\phi}{\cos\phi} = \tan\phi = \sqrt{3}\dfrac{W_B - W_A}{W_B + W_A},$

from which relation ϕ and $\cos\phi$ (power factor) may be obtained.

The writer has found this method of determining the power factor of circuits of great use in experimental work.

In all diagrams given in relation to power measurement the wattmeters have been shown in direct connection with the circuit but, in practice, in high tension circuits, instrument transformers will almost invariably be used both for the current and pressure coils, and also in connection with the current coils when heavy currents are in use on low pressure circuits (see page 116).

EXAMPLES

1. Determine the numerical relationship between the phase and line pressures in the case of a star connected four phase system.

Answer. $1 : \sqrt{2}$.

2. Determine the numerical relationship between phase and line currents in the case of a six phase mesh connected system. *Answer.* $1 : 1$.

3. Construct a diagram to show that if the ninth harmonic is present in the wave form of a mesh connected three phase system, it will give rise to current circulating round the mesh.

4. Using equal current densities in all the lines, compare the amounts of copper used in transmitting a certain amount of energy by the single phase and the two phase three wire systems. Arrange for the same maximum pressure between any two lines in each case. *Answer.* $1 : 1\cdot21$.

5. Using a pressure of 20,000 volts between lines in each case, compare the amounts of copper necessary to transmit 1000 K.W. (*a*) by a single phase system, (*b*) by a three phase star connected system, equal current densities being used in each case. *Answer.* $1 : \cdot867$.

6. Repeat the last example on the assumption that equal losses are to be arranged for instead of equal current densities. *Answer.* $1 : \cdot75$.

7. Demonstrate, by using the method of instantaneous values, that two wattmeters, connected as in Fig. 63, will indicate the true power in a three phase star connected system under any conditions of load.

8. Repeat the last example in the case of a mesh connected system. (In this case when a winding is passed through clockwise take the current and pressure as positive and *vice versâ*.)

9. Draw a complete vector diagram, showing line and phase pressures and currents, for a star connected three phase system in which the phase current lags 45° behind the phase pressure. (Assume the load to be balanced.)

10. Repeat the last example for a balanced three phase mesh connected system in which the phase current is 20° in front of the phase pressure.

CHAPTER VI

Ammeters and Voltmeters.

It is of the utmost importance that all devices made use of for the measurement of current and pressure in alternating current circuits should be so arranged that, for any one position of the moving part, the force causing the deflection, or the tendency to deflect, shall be proportional to the square of the current or pressure as the case may be. If this is so the average force causing deflection will be proportional to the average square of the current or pressure during one complete cycle, and then, by suitably calibrating the scale, we can make the instrument read the square root of mean square value of the quantity concerned. It follows from the above statement that the type of scale usually obtained on alternating current ammeters and voltmeters is *likely* to follow the square law, that is, doubling the current will cause four times the deflection; we may also note that in all such cases the direction of the deflection will be independent of the direction of the current. Departures from this square law scale are due to the movement of the moving part affecting the forces causing the deflection quite apart from any variation of the current; the type of control may also influence the scale obtained. Hot wire type instruments give good square law scales, but moving iron type instruments have scales which are much affected by the causes mentioned above.

All the common types of alternating current ammeters and voltmeters, apart from those specially designed to follow the variations during a cycle, have usually sufficient inertia (either mechanical or thermal or both) to ensure a steady reading being obtained; if this were not so the pointer would be in a continual state of vibration and indeed this is occasionally observed in instruments of the moving iron type.

Moving iron type instruments. A very common type of instrument for use on alternating current circuits is that in which we have a solenoid containing a movable piece of iron to which is attached the pointer; when current passes round the coil the iron becomes magnetised by induction and the force developed on this piece of iron by the magnetic field causes the required deflection. The soft iron in such instruments is worked at a very low flux density, and, under this condition, the intensity of magnetisation produced in the iron is practically proportional to the magnetising force, so that when the current round the coil is doubled the magnetic field and the magnetisation of the iron are both doubled thus quadrupling the deflecting force; generalising, we see that in such cases the deflecting force will be proportional to the square of the current. As a matter of fact the magnetisation of the iron is only approximately proportional to the current and so the square law is only approximately true, and this fact results in the indications of such instruments depending to some extent upon the wave form of the current passing round the coil. Both ammeters and voltmeters are constructed on this principle, the coils of the former consisting of a few turns of stout wire and of the latter of many turns of fine wire. In the case of the voltmeter, therefore, the coil will have a certain amount of inductance, in addition to the resistance, and the indication of the instrument is likely to depend to some slight extent on the frequency of the supply, for, if the frequency is increased, the reactance, and consequently the impedance, of the circuit will rise resulting in the same pressure sending less current through the instrument and consequently giving less deflection; a diminution of frequency will cause the opposite result. Again, if the coil of the voltmeter is, as is usually the case, wound with copper, the resistance of the coil will depend upon the temperature and thus the indications of the instrument will also depend upon the temperature. Moving iron type instruments are cheap, robust and not readily burnt out and consequently of great service where extreme accuracy is not essential. Care should always be taken to avoid the risk of the

7—2

production of any considerable eddy currents in metallic formers
or other metal parts adjacent to the coil as this would result in
inaccuracy, especially in ammeters; if a metallic former is used
to carry the coil it should always be split longitudinally. Many
types of this class of instrument are met with in practice, differing
in details of construction though not in principle, and, since
they have been so frequently described, it has not been considered
necessary to enter into detail concerning them in the present
work.   In conclusion it may be stated that either gravity or
spring control may be used and that air damping, usually not
very efficient, is commonly resorted to.

Hot wire type instruments. Another device employed with
great success in alternating current ammeters and voltmeters is
the expansion of a wire through which a current, proportional to
the current or pressure to be measured, is passed.

If the current passing through a wire of constant resistance is
doubled the rate of production of heat will be quadrupled; the
expansion of the wire will be proportional to the temperature rise
of the wire and this, in turn, will depend upon the average rate of
production of heat during the cycle; the expansion of the wire
will therefore be proportional to the average value of the square
of the current during the cycle and thus, by measuring the expansion
and suitably calibrating the scale, we may construct an instrument
suitable for indicating the R.M.S. value of a current or a pressure.

In the earlier Cardew type of hot wire instrument the direct
expansion of the wire was measured, but in more modern types
the expansion of the wire is allowed to cause a sag which is taken
up and measured by a suitable arrangement; this form of instrument is very compact and much more convenient than the older
type. Instruments on this principle, which can be used in conjunction with shunt and series resistances as ammeters and voltmeters respectively, give indications which are, as far as all
practical purposes are concerned, independent of frequency and
wave form and hence are particularly useful for general testing
purposes where considerable variations in those two factors are
to be expected. Their chief weakness is the ease with which they
are burnt out, this being effected by a comparatively small excess
of the full load current. They are quite inexpensive.

The Hartmann and Braun pattern of hot wire instrument, as made by Messrs Johnson and Phillips, is illustrated in Fig. 65, which shows an ammeter arranged to indicate a maximum current of two amperes directly and of course higher currents with the help of suitable shunts. The hot wire AB is composed of Platinum Silver, about ·33 mm. in diameter, and is stretched between two clamping pillars A and B. Current is fed into the wire at the centre and taken out at the ends, thus the two sides of the wire are in parallel so far as the flow of current is concerned, this allows of the use of a smaller size of wire. When current passes, the wire AB is caused to sag downwards and this, in turn, causes the phosphor bronze wire CD to sag. The wire CD is connected to the spring E (which is always pulling to the left) by the silk fibre JK which passes round, and is firmly secured to, the small pulley carrying the pointer N; we see, therefore, that the sag of AB is finally taken up by the spring and results in an appropriate movement of the pointer from left to right. These instruments are apt to show an alteration of their zero point from time to time, especially if subject to overloads, and a simple adjustment is provided by the set screw SC working through the collar Z into the lever L. A damping device is provided comprising an aluminium disc Al rotating between the poles of a permanent magnet PM; in addition to assisting in bringing the pointer rapidly to rest, this device is also important in so far as it prevents violent oscillation of the pointer, giving rise to stresses on the wires, during transport of the instrument. As will be seen from the diagram the whole movement is mounted on a suitable base the material of which is chosen so as to have a coefficient of expansion with temperature rise as nearly as possible equal to that of the wires, this will tend to prevent alteration of temperature of the instrument, as a whole, affecting the reading. The voltmeter movement is constructed on essentially the same principle but a much finer hot wire is used (about ·06 mm. in diameter) and the current, which is of course quite small, is sent through the wire from end to end.

Another type of instrument depending upon the heating effect of the current is the Duddell Thermo-ammeter which is made by the Cambridge Scientific Instrument Co. This instrument (see Fig. 66)

comprises a moving coil permanent magnet type movement, similar
to those used for direct current purposes, but the ends of the coil,
instead of being brought out externally by means of the springs,
are connected directly to the ends of a small BiSb thermo-couple;
this couple revolves with the coil and is situated immediately above
a small heater through which the current to be measured is passed.
The effect of the passage of current through the heater resistance

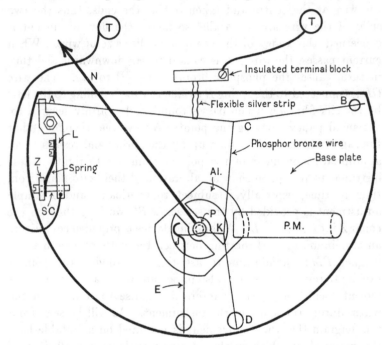

Fig. 65. Diagram of hot wire movement (balance weights not shown).

causes the development of heat, which is transmitted to the thermo-
couple, thus causing the production of a current round the coil
which in turn causes a deflection of the spot of light or the pointer.
The heater is composed of platinised mica for small currents and
of wire for larger currents and in either case is practically devoid
of capacity and inductance, hence the instrument is very
suitable for the measurement of currents and pressures of high
frequency such as are met with in telegraphy, telephony and
wireless work.

Ammeters on this principle have been constructed for currents as high as 10 amperes but they are essentially suitable for small currents, of the order of a few milli-amperes, and a low reading ammeter can readily be converted into a voltmeter with the help of a suitable series resistance.

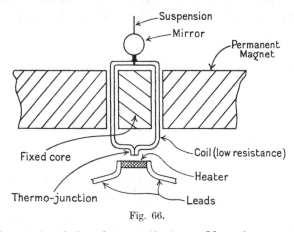

Fig. 66.

Instruments of the electrostatic type. Many instruments, including ammeters, voltmeters, wattmeters and ohmmeters, have been constructed on this principle but the small deflecting forces produced, necessitating correspondingly small controlling forces, lead to the production of a delicate type of instrument and, so far as commercial alternating current instruments are concerned, only voltmeters need be considered. Apart from the difficulty mentioned, the type has many important advantages and has extensive use in laboratory and test room work. The indications are independent of temperature, wave form, frequency and external magnetic fields, and the power taken by voltmeters is negligible. On direct current circuits voltmeters allow no current to pass through them, if the insulation is perfect, except at the instant of switching on, but on alternating current circuits a small current passes on account of the electrostatic capacity, which is of course an inherent feature of their construction; this current is practically wattless.

For low and medium pressure electrostatic voltmeters the general mode of construction and operation is similar to that of the quadrant electrometer, though the electrical connections are usually not quite the same. One of the best known types of instrument is that made by Messrs Kelvin, Bottomly and Baird in which there are two essential parts—a pair of fixed metallic cells A which are connected to one pole of the pressure to be measured, and a metallic vane B, which is suspended and controlled by a quartz or metallic fibre, and which is connected to the other pole.

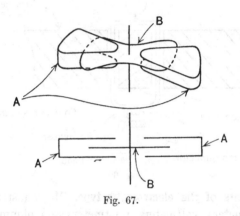

Fig. 67.

When in its zero position the moving vane is quite clear of the cells, but, when a pressure is applied to the instrument, forces are developed tending to draw the vanes into the cells, the former moving until the forces of attraction (due to ordinary electrostatic principles) are balanced by the opposing forces of the control. When dealing with low pressures the force developed per cell is very small and so, for pressures up to 600 volts, a number of cells (about 10) are arranged under each other, the corresponding vanes, which turn about a vertical axis, being rigidly connected together, this arrangement giving what is commonly known as the multicellular type. The pointer, which is connected to the vanes, is usually rather sluggish in its movements and the necessary small amount of damping is effected by a small disc or wire attached to the lower end of the vanes and moving in oil of proper viscosity. The zero of the instrument can be adjusted by first levelling the instrument, with the help of the small spirit level provided, and then (if necessary) by means of the milled head, care being taken that no difference of pressure exists between the cells and vanes during this process.

An instrument of this type, reading up to say 120 volts when connected directly on to the circuit, can be used for higher pressures if, in connection with the instrument, we use a resistance multiplier. A diagram of connections of this device is shown in Fig. 68, from which it will be seen that the line pressure is connected across the complete resistance and the instrument is connected across a part of the resistance such that the actual pressure on the instrument never exceeds 120 volts. Thus, for instance, a 120 volt instrument can be fitted with a multiplier so that full scale deflections can be obtained with pressures of 120, 240, 480 and 600 volts. When

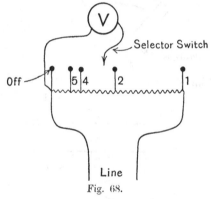

Fig. 68.

the multiplier is in use the combination is no longer electrostatic, since a conductance current will be passing through the multiplier; the other good qualities of the type will, however, be retained.

For medium pressures, up to say 20,000 volts, the same general type of construction is adopted but only one pair of cells is made use of and the vanes move about a horizontal axis. Instruments of this type are usually gravity controlled, and the actual range of any one instrument can be modified by using one or other of the several control weights supplied with the instrument. It is very important in this class of instruments (and in those for higher pressures) that adequate sparking distances be provided between those parts of the instrument which are oppositely electrified and that stops be employed to prevent the vanes and cells from coming into contact.

For still higher pressures, the principle of direct attraction between surfaces at different potentials is employed, the general arrangements being as indicated in Fig. 69, in which one pole of the pressure to be measured is connected to the lower stationary electrode and the other pole to the upper movable electrode (and also to the guard ring C). When pressure is applied to the instrument, attraction takes place between the two electrodes and the pointer moves over the scale until the attracting force is balanced by the counter force exerted by the control weight, by altering which, the range of any one instrument can be modified.

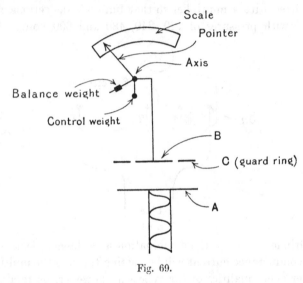

Fig. 69.

Since the attraction between the electrodes is proportional to the square of the difference of pressure between them, a necessary condition for accurate results on alternating pressures, if we wish to double the range of the instrument we must quadruple the control weight.

Dynamometer type ammeters and voltmeters. An instrument of this type comprises two coils, one fixed, and the other pivoted, spring controlled and of course capable of rotation round the line joining the pivots as an axis. The moving coil is in the region occupied by the magnetic field produced by the fixed coil and *vice versâ*. When the same current passes through both coils in series a force develops between them which, for any one position of the coil, is proportional to the square of the current, and the coil which is free to move takes up a deflection such that the average deflecting force is balanced by the counter force due to the control springs; it will be clear, from a consideration of the general principles already enunciated concerning alternating current ammeters and voltmeters, that the deflection will be proportional to the R.M.S. value of the current concerned. The general arrangements of such an instrument are very similar to those in a permanent magnet moving coil instrument as used on D.C. circuits with the important exception that the magnetic field, produced by the

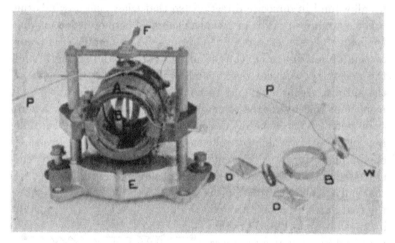

Fig. 70.

permanent magnet in the latter case, is replaced by the magnetic field produced by the fixed coil. The dynamometer type of instrument is available for use as an ammeter, a voltmeter or a wattmeter, differences only existing in the connections and arrangement of the coils. The general arrangements of a movement of this type, as made by Messrs Elliott and Co., Ltd. (to whom, and to Mr W. Phillips, M.I.E.E., A.C.G.I., the author is indebted for information and diagrams), are shown in Fig. 70, which illustrates a movement suitable for currents up to one ampere. The fixed coil A consists of a fairly large number of turns of insulated wire and the moving coil B, which is made as light as possible, consists of one layer of

silk covered aluminium wire the ends of which are soldered to
small brass stampings to which are also soldered the ends of the
control springs, thus securing a continuous metallic circuit through
the moving coil. The moving coil is supported on, but insulated
from, a light aluminium staff which also carries, at its lower end,
two light aluminium vanes D which are caused to damp the
instrument by rotation in a box E arranged to give but little
clearance. Two small hardened steel pivots are let into the ends
of the staff and these are pivoted into two jewels of polished
sapphire carried in a suitable manner by the frame of the instrument. The control springs consist of special bronze and are, as
usual, wound in opposite directions so that when one unwinds the
other is wound up; this prevents a change of zero or deflection due
to change of temperature of the springs. The pointer P is made
of thin aluminium tube, thus securing lightness and strength, and
the necessary balance weight is shown at W. The free ends of
the springs are brought out to connecting plates, the upper one
being movable by means of the lever F in order to secure a zero
adjustment. The type of movement shown will also be suitable
for voltmeters in which case the current will of course be less than
one ampere; for larger currents than one ampere the total current
is passed round the fixed coil but the moving coil is shunted so
that only a definite fraction of
the toal current passes through
it, as shown in Fig. 71; this
modification of course produces
no change in the law of the
instrument. The resistance R
shown in this figure has a
threefold use, in the first place
it is convenient as an adjusting
resistance; it also, since it is
composed of material having

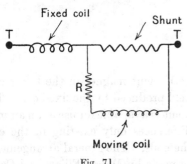

Fig. 71.

a negligible temperature coefficient, serves to diminish the
effect of temperature on the calibration of the instrument which
otherwise might be undesirably large due to variation in resistance
of the moving coil; finally, it keeps down the change of impedance
of the moving coil circuit due to change of frequency.

A standard full load current for this type of ammeter is 5 amperes which is very suitable for use in connection with current transformers on high pressure circuits, the scale of the instrument of course being calibrated to read directly the current in the main circuit.

When used as a voltmeter both coils are wound with fine wire and it is desirable to have the movement so sensitive as to allow of the introduction, in series with the coils, of a comparatively large non-inductive resistance of wire having a negligible temperature coefficient so as to reduce the effect on the calibration of changes of frequency and temperature; this can be more readily effected on the higher pressure ranges. Instruments of the dynamometer type have long been used under the names of Siemens' Dynamometer and Kelvin's Balance, but these, though presenting the possibility of great accuracy, suffer under the disadvantage of being zero type instruments, that is the pointer must be brought to zero by some suitable adjustment before the reading can be taken, and this militates against their use when readings must be taken quickly, especially when the current or pressure is at all unsteady.

Dynamometer instruments having iron cores. In instruments of the moving coil dynamometer class the forces tending to cause deflection are small because it is not easily possible to produce strong magnetic fields, with such coils as may be used in instruments, without the use of an iron core. In general the tendency of makers of alternating current instruments has been to avoid the use of iron when a fair degree of accuracy was required. Professor Sumpner, however, has introduced a series of instruments (including ammeters, voltmeters, wattmeters, etc.) in which an iron cored electromagnet is employed, thereby obtaining the strong deflecting forces which are so advantageous both on portable and switchboard instruments. One of the most interesting, and, at the same time most useful, of these instruments is the voltmeter the connections of which are shown in Fig. 72, from which it will be seen that there are two circuits through the instrument; the first is the exciting circuit of the shunt electromagnet which produces in the gap a flux proportional to the pressure and practically 90° behind it in phase; the second is the circuit containing the moving

coil which is in series with a condenser and, since the resistance of
this circuit will be comparatively small compared with the condensance, we shall have the current in the coil leading the pressure
by practically 90°. Thus the maximum current in the coil will
occur at practically the same instant as the maximum flux in the
gap and the conditions are favourable for the production of a
considerable torque. The actual deflecting force will be co-jointly
proportional to the current in the coil and to the flux density in
the gap and, since each of these is proportional to the pressure,
the deflecting force will be proportional to the square of the
pressure and this is the condition which we have seen must be
satisfied in all alternating current voltmeters. The coil is controlled in the usual manner by two springs which also serve to
convey the current to the moving coil.

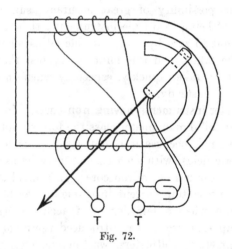

Fig. 72.

In connection with this instrument two features are of especial
interest from the more theoretical point of view: in the first place,
the flux in the gap is settled by what may be termed choking coil
principles, that is it takes up such a value that the induced back
pressure is practically equal (numerically) to the applied pressure
(see page 41), and so, for a given value of the applied pressure, is
independent of the nature of the iron and of the exact length of
the gap; it will also be proportional to the voltage applied. In
the second place, it is interesting to note that the effect of change

of frequency on the indications of the instrument is very small since doubling the frequency will, if the pressure is constant, halve the gap flux and double the current through the coil, thus leaving the force causing the deflection unaltered. Ammeters have also been constructed on similar principles but in this case it is necessary to use a series excited magnet, and the flux produced will then depend upon the permeability of the iron, which may vary with time, and upon the gap reluctance.

Induction type ammeters and voltmeters. A number of types of instruments have been developed depending upon the production of induced currents in masses of copper or aluminium suitably

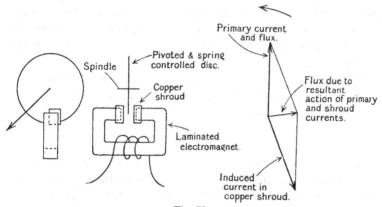

Fig. 73.

proportioned and situated in an alternating magnetic field. One of the most common arrangements, so far as ammeters and voltmeters are concerned, consists of a copper or aluminium disc which is pivoted, spring controlled and situated so that its edge is within the strong magnetic field produced by a laminated electromagnet suitably excited. One side of each pole of the electromagnet is covered or shrouded by a piece of copper as indicated in Fig. 73.

Perhaps the simplest explanation of the mode of action is arrived at by considering the magnetic fields produced by the main current and the induced current in the copper shroud. The field which cuts the copper disc to the left of the centre line of the electromagnet may be looked upon as being produced by the current in the

coil alone (*i.e.* by what may be termed the primary current),
while the field which cuts the copper disc to the right of the centre
line will be the resultant effect of the primary current and of the
current which is induced in the copper shroud. The current in
the shroud will lag behind the primary current by an angle of
about 150° and the resultant field will therefore be some 90°
behind the primary current, that is, behind the flux to the left of
the centre line*.

The general effect will therefore be a progress of the flux from
left to right tending to cause, for the case shown in the diagram,
an anti-clockwise motion of the disc. In this simple form the
instrument will be much affected by change of frequency because
the magnitude of the current induced in the disc will depend upon
this factor (especially in ammeters). Change of temperature will
also affect the calibration, chiefly because of the alteration of the
specific resistance of the disc.

This type of instrument is more suitable for switchboard work,
where frequency and wave form are practically constant, than for
general testing purposes.

Wattmeters.

In these instruments, contrary to what is the case in ammeters
and voltmeters, we must use a movement such that the deflecting
force, for any one position of the moving part, is proportional to
the instantaneous value of the power; the average force for one
position of the movement will then be proportional to the average
value of the power during the cycle and it is the average value of
the power that we need to measure (see page 62).

The direction in which the moving part of the wattmeter tends
to deflect at any instant depends on the sign of the power at that
instant, and, in most practical cases, the power will reverse four
times during the cycle, owing, however, to the considerable inertia
of the moving portion, a steady deflection will be obtained which
correctly indicates the true average power. Since the instantaneous
value of the power depends on the value of the product of the
instantaneous values of pressure and current, each wattmeter must
have two coils; one, the current coil, carrying a current which is

* See page 193.

proportional to the current in the circuit in which the power is to be measured; and the other, the pressure coil, carrying a current which is proportional to the pressure of that circuit*.

In the usual dynamometer type (see page 107) of wattmeter the current coil is most commonly fixed and the pressure coil, to which the pointer is attached, is pivoted so as to move within the current coil; there is, however, nothing to prevent the reverse arrangement being adopted if desired.

Now the average value of the power in a circuit is equal to the average value of (current × pressure) during a complete cycle, and the reading of the instrument will be proportional to the average value of (current in current coil × current in pressure coil); it is of the utmost importance to realise that not only must these two currents be proportional to the current and pressure in the main circuit respectively, but the phase difference between the currents in the current and pressure coils must be the same as the phase difference between the current and pressure in the main circuit; in practice the latter condition is the more difficult to arrange for. Suppose, in a dynamometer type of wattmeter, the full current is allowed to pass through the current coil, there is then evidently no scope for phase error, as it is called, in that portion of the wattmeter; the pressure coil, however, will consist of many turns of wire (having some slight inductance) connected in series with a non-inductive resistance and obviously the current in this coil will lag by some small angle behind the pressure. Thus, if the load is an inductive one, the phase difference between the two currents in the instrument will be rather less than the phase difference between current and pressure in the main circuit and the instrument will therefore read slightly high; on the other hand, if the current leads the pressure in the main circuit, the instrument will read a little too low. Vector diagrams illustrating these points are given in Fig. 74 in which ϕ represents the phase difference between current and pressure in the external circuit, and ϕ_1 represents the phase difference between the current in the current coil and the current in the pressure coil of the wattmeter.

As a matter of fact the phase error is not likely to be at all

---

* This statement does not apply to the electrostatic type of wattmeter which, however, is too delicate for ordinary commercial use.

serious in the type of wattmeter under consideration if the instrument is directly connected to the circuit under test, but, if instrument transformers are used, as on high pressure or high current
circuits, the error is likely to be far more serious.   In any one
instrument the phase error is likely to be more serious on the low
pressure ranges, since the reactance of the pressure circuit will
then bear a greater proportion to the resistance of that circuit,
thus giving a greater phase difference between the pressure and
the current in the pressure coil.

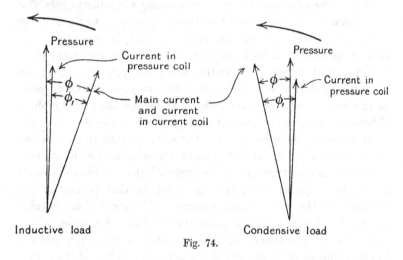

Inductive load   Condensive load

Fig. 74.

The phase error of an instrument will also exert a greater effect
on the accuracy of the instrument when the power factor of the
main circuit is low; thus, for the sake of illustrating the point,
assume the phase error is ·5° (a far larger value than will exist in
a dynamometer instrument even on low pressure ranges), the
pressure 100 volts and the current 10 amperes and let us investigate
the error as we vary the power factor.

The table clearly shows the increasing importance of the phase
error as the power factor of the circuit becomes lower, and that
the wattmeter would actually read backwards if placed on a
circuit in which the current was leading by nearly 90°.   As stated
previously the phase error in commercial dynamometer wattmeters
is considerably less than that assumed in the foregoing example

and the author has been informed by Mr Phillips that in a watt-meter made by Messrs Elliott for use on a 100 volt circuit, the inductance of the pressure circuit was ·0017 henry, the resistance being 7540 ohms; in this case at 50 cycles per second the tangent of the angle of phase error is ·00007 and the angle itself is about ·004°, the error introduced in the reading of the wattmeter only being ·04 % even when the power factor is as low as ·17, this is a quite negligible amount.

Angle of lag in external circuit ϕ	$\cos \phi$	True watts	Angle between currents in instrument ϕ_1	$\cos \phi_1$	Reading on instrument	Error		
						Watts	Per cent.	
Current lagging								
0°	1·	1000·0	·5°	·9999	999·9	·1	·01	Low
30°	·866	866·0	29·5°	·8704	870·4	4·4	·5	
60°	·500	500·0	59·5°	·5075	507·5	7·5	1·5	High
90°	·000	0·0	89·5°	·0087	8·7	8·7	∞	
Current leading								
0°	1·	1000·0	·5°	·9999	999·9	·1	·01	
30°	·866	866·0	30·5°	·8616	861·6	4·4	·5	Low
60°	·500	500·0	60·5°	·4924	492·4	7·6	1·5	
90°	·000	0·0	90·5°	·0087	8·7	8·7	∞	High*

Another trouble which may cause error in wattmeters is the production of eddy currents in parts adjacent to the movement; these might alter the phase and magnitude of the magnetic fields produced by the coils of the instrument. The general mechanical arrangements of a dynamometer wattmeter will be similar to those of the instrument shown in Fig. 70 with the exception that the fixed coil would generally consist of fewer turns of stouter wire and the fixed and moving coils would no longer be connected in series.

Instrument Transformers.

In many cases instruments on alternating current circuits are connected up through small transformers, known as instrument

* Wattmeter reading in opposite direction.

transformers, instead of being connected directly to the circuit. There are two distinct reasons for this procedure; in the first place, on high tension circuits, very careful insulation would be required in instruments if connections from the high pressure lines were taken directly into them, the use of instrument transformers clearly avoids the necessity of doing this; secondly, when dealing with heavy currents whether at high or low tension, constructional difficulties appear in connection with the current coil and these again may be avoided with the help of current transformers if the secondary is arranged to pass a current considerably less than, but proportional to, the main current.

Instrument transformers may be divided into two classes:

(1)   Those intended for use in connection with voltmeters and volt coils of other instruments. These are known as pressure transformers.

(2)   Those intended for use with ammeters and with current coils of other instruments. These are known as current transformers.

All instrument transformers should fulfil as far as possible the following conditions:

(1)   The magnitude of the secondary pressure (in pressure transformers) or current (in current transformers) should be proportional to the primary pressure or current as the case may be.

(2)   The secondary pressure (or current) should, as regards phase, be in direct opposition to the primary pressure (or current).

(3)   There must be very good insulation between primary and secondary circuits, especially when they are used for high tension systems.

Requirements (1) and (3) must be met in all instrument transformers but (2) is only of importance when the transformers are to be used in connection with wattmeters, supply meters and other cases where the phase relation between pressure and current is of importance.

Pressure transformers. These consist of a laminated core of the closed circuit type provided with primary and secondary windings, the primary winding having many more turns than the secondary. If the resistance of the windings, the magnetic leakage, magnetising current and core loss are all kept as small as possible the secondary pressure will be very nearly proportional, and in phase opposition, to the primary pressure.

Current transformers. Here again we have a laminated iron core of the closed circuit type. The primary winding now consists of a few turns only and in fact in many cases the straight conductor is simply passed through the core (as in Fig. 75). The secondary circuit will comprise a greater number of turns, the exact number depending upon circumstances, but in any case both the pressure and current in the secondary will be fairly low, a common value for the full load secondary current being 5 amperes. Usually

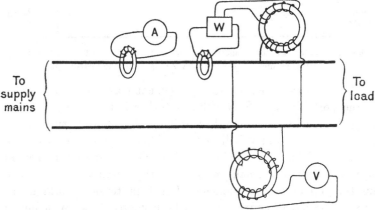

Fig. 75.   Connection of ammeter, voltmeter and wattmeter to high pressure lines when separate transformers are used for each instrument.

when dealing with transformers we have an impressed pressure on the primary (*i.e.* the line pressure) which is constant at all loads but in the case of the current transformer, where the primary is in series with the load, this is not the case, the applied pressure to the primary being low and of course variable with load.

Despite this fact the ordinary vector diagram of the transformer applies if the magnitudes of the various quantities are suitably modified. If the magnetising current of the transformer is kept low, the secondary current will be practically proportional to the primary current and 180° from it in phase (except for very small loads).

Separate transformers may be used for each indicating instrument if desired and perhaps greater accuracy will be obtained by so doing but it is permissible, in order to limit the number of transformers, to use one current transformer to operate an ammeter and the current coil of a wattmeter (the two coils being connected in series as in Fig. 76), and a single pressure transformer to operate a voltmeter and the pressure coil of a wattmeter (the two coils being in parallel as in Fig. 76).

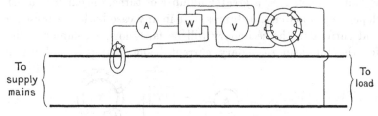

Fig. 76. Connection of ammeter, voltmeter and wattmeter to high pressure circuit when one transformer serves more than one instrument.

A view showing the general arrangement of a current transformer of the portable type (for which the author is indebted to Messrs Everett, Edgcumbe and Co., Ltd.) intended for a low pressure circuit is shown in Fig. 77; it will be noticed that the terminals for the secondary do not appear in the view shown but as a matter of fact they consist of small terminals of the clamp variety situated on the top of the instrument. For high tension working the primary would of course need to be insulated to a much higher degree than in the case shown*.

In conclusion it may be stated that relays and trip coils on high tension circuits are also frequently operated by small transformers such as are dealt with above, and in certain cases the same transformers may be used to operate both instruments and relays.

Frequency Meters.

The exact determination of frequency is a matter of considerable importance both in central station work, where it is necessary that the frequency be kept constant, and in general testing work, where frequencies varying over a wide range are required to be

---

* Readers desiring a fuller knowledge of instrument transformers should consult a paper by Mr Kenelm Edgcumbe in the *Electrical Review*, Vol. 67, p. 163.

determined. When the alternator producing the current is accessible the frequency can be readily determined from a knowledge of the speed and the number of poles, but even in this case it is more convenient to use a direct reading instrument, and in many instances it is essential to do so. Modern commercial

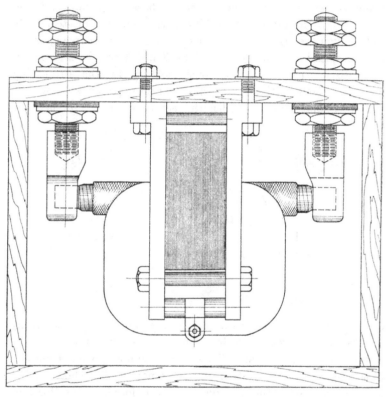

Fig. 77. Elevation of current transformer for low pressure circuit.

indicators are usually founded on one or other of the following principles:

(1) The production of resonance in steel reeds, or

(2) The variation of current in an inductive circuit.

Frequency indicators of the resonance type. This type was originally suggested by the late Professor Ayrton and has been developed by, amongst others, Messrs Everett, Edgcumbe and

Co., Ltd., to whom the author is indebted for information on the matter. The principle and general arrangement of one pattern of instrument is shown in Fig. 78 in which A represents a central laminated iron core to one end of which is attached a circular flat iron plate B; the core is surrounded by a coil which is excited by being connected across the pressure whose frequency is to be measured. The brass support D carries a number of steel reeds as C, which are arranged in a circle and so that their free ends are in close proximity to the iron plate B which will be alternately magnetised with north and south polarity by the alternating current. Each of these reeds has a natural period of vibration (which is dependent on the length, section, etc.) and, further, the

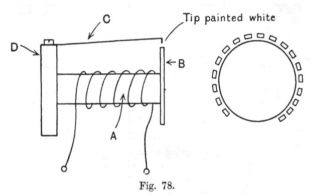

Fig. 78.

alternating magnetism in the iron plate impresses upon the reeds an alternating force whose frequency will be twice the frequency of the pressure under test. These impressed forces will not succeed in setting a reed in vibration unless the natural periodicity of the reed is very near the periodicity of the force. Thus, for any periodicity, the reed whose natural periodicity is nearest to that of the impressed force will be set in vibration, and this particular reed can readily be seen from the front of the instrument owing to the apparent broadening of the edge which is painted white so as to render observation easy.

Usually, in practice, two or more reeds are set in vibration and, by noting the relative amplitudes of the vibrations of the reeds concerned, the frequency can be estimated with a considerable degree of accuracy.

Each instrument is fitted with a number of reeds (varying in different types from 15 to 55), each reed having a slightly different period from its neighbours, the exact difference depending upon the number of reeds and upon the range of frequency which the instrument is designed to cover. For central station working, where the range of frequency variation is small, successive reeds may differ in frequency by a quarter of a period per second, and this will usually allow the frequency to be estimated to about one-tenth of a cycle per second; for general test room work a coarser gradation will be advisable. It is interesting to note that

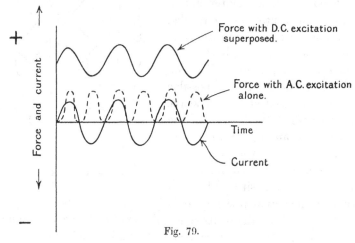

Fig. 79.

if the core is provided with an additional winding, through which a direct current is passed of such a strength as to cause the resultant flux in the core to pulsate in one direction only rather than to alternate (see Fig. 79), there will be but one period of maximum force per cycle, instead of two as in the former case, and thus the frequency indicated by each reed will be doubled. Thus an instrument which reads from 30 to 50 cycles per second in half cycle steps when the alternating magnetising force is alone used, will, when the direct magnetising force is superposed, read from 60 to 100 cycles per second in steps of one cycle.

Frequency meter depending upon the variation of current in an inductive circuit. An efficient frequency meter may also be constructed by taking advantage of the fact that any alteration of the frequency of the pressure applied to an inductive circuit, all the other factors remaining constant, causes an alteration in the magnitude of the current.   In the instrument working on this principle made by the Weston Electrical Instrument Co., Ltd. (to whom the author is indebted for information), there are two fixed coils, C_1 and C_2, connected in combination with reactances L_1, L_2 and L_3, and resistances R_1 and R_2, as shown in Fig. 80. The pivoted moving system comprises a thin piece of iron (moving within the fixed coils), a pointer and a damper, and is entirely uncontrolled by springs so that, apart from the influence of the fixed coils, it is free to take up any position.   The iron is chosen to be as free as possible from hysteresis effects and its shape precludes the possibility of the formation of any appreciable eddy current; each of the fixed coils will therefore tend to make the iron take up a position along its own axis, the actual position taken

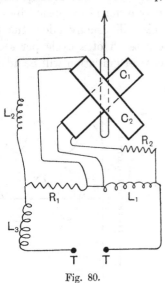

Fig. 80.

up depending upon the relative magnitude of the currents through the two coils.   At normal frequency the coils are adjusted so that the iron takes up a position bisecting the angle between the axes of the coils; if the frequency increases the current in C_2 falls while that in C_1 rises thus causing the pointer to move in an anti-clock-wise direction, a fall in frequency of course producing an opposite effect.   The scale of this type of instrument is obviously perfectly continuous and the indication is but little affected by alteration of pressure or wave form.

Power Factor Indicators.

The power factor of a circuit can be readily found from a knowledge of the volt-amperes and the watts, being given by the ratio $\dfrac{\text{watts}}{\text{volt-amperes}}$, but this method necessitates three observations

and a subsequent calculation and is obviously very inconvenient
for switchboard working and, to a less extent, for general testing
work, and these considerations have led to the introduction of
direct reading power factor meters.   The construction of such an
instrument suitable for a single phase circuit is shown in Fig. 81,
where C is a fixed coil of stout wire which is traversed by the
main current, and R and X are two coils, rigidly connected with
their axes at right angles, the combination being furnished with
a pointer and pivoted; since the moving coils are not controlled
by springs they are free, apart from any force exerted by the
fixed coil, to take up any position.   From the diagram it will be

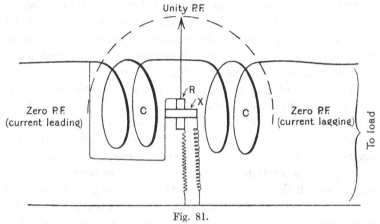

Fig. 81.

seen that X is a pressure coil in series with reactance and that
R is also a pressure coil but in series with resistance, the current
in each case being led into the coils by strips which exert practically
no controlling force.

In order to study the action of the instrument suppose for the
moment that the load is purely inductive in which case the currents
in coils C and R will differ in phase by 90° and so there will be no
resultant force between them; on the other hand the currents in
coils C and X will be in phase with each other and the coil X will
therefore tend to set itself parallel with the coil C, that is the
pointer will take up a position at right angles to that shown in
the diagram and this point should be labelled zero power factor
or 90° lag.   Again, if the load is next considered to be purely

non-inductive, the currents in C and X will differ in phase by 90° and so there will be no resultant force between these two coils; but the currents in coils C and R will be in phase, the coil R will set itself parallel to the coil C, and the pointer will take up the position shown in the diagram which should be labelled unity power factor or zero angle of lag. The direction in which the pointer will deflect in the former case will depend upon the direction of the windings but it will be quite clear that a lagging current will cause a deflection in one direction and a leading current in the opposite direction. For cases usually met with in practice the phase difference between pressure and current will be between 0° and 90° and under these conditions force will be exerted on both coils and both R and X will tend to set themselves parallel to C, the final position taken up depending upon the relative forces exerted by the coils which clearly depend upon the phase angle between current and pressure in the main circuit.

Another point of view from which the action of the instrument may be appreciated is obtained by considering the action of the two phase currents (obtained by the phase splitting device) in the circuits of the moving coils. These currents will produce a rotating magnetic field (see page 226) and the moving coils will always take up such a position that the line of action of the rotating field coincides with the line of action of the alternating field (produced by the fixed coil) at the instant the latter is passing through its maximum value. If the particular cases of the purely inductive and non-inductive loads dealt with above be considered in the light of this point of view it will be seen that precisely the same positions of the pointer will be arrived at as in the consideration above. When considering a single phase power factor meter the first point of view seems somewhat simpler but, when considering two and three phase power factor indicators, the second method is more helpful.

Single phase indicators will need to be calibrated for a particular frequency, since this factor will materially influence the current through the coil X, and small errors may also be introduced into the readings by variations in the wave form and magnitude of the applied pressure.

In the case of two and three phase power factor indicators,

for balanced loads, the general arrangements are very similar but no phase splitting device is needed, the required difference in phase between the currents in the moving coils being obtained by connecting them across the different phases of the supply circuit. The diagram of connections for the three phase instrument is shown in Fig. 82 (a), and it will be clear that the instrument really indicates the power factor of the particular line in which the

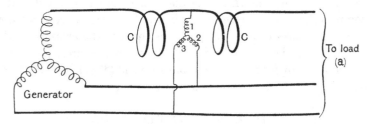

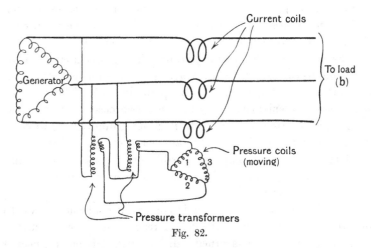

Fig. 82.

current coil is inserted. The three moving pressure coils are connected so that the angle between the axes of any two of them is 120°.

If the load is unbalanced three fixed current coils are necessary, these also being arranged so that their axes are 120° apart, and the connections of such an instrument are shown in Fig. 82 (b). The moving system will in this instance take up a position so that the directions of the axes of the two rotating fields (produced by the current coils and the pressure coils respectively) will, at each instant, be coincident*, the instrument thus indicating the average power factor of the three phases.

Fig. 82 (*b*) is of particular interest since it illustrates the feeding of the three pressure coils from two pressure transformers, a similar arrangement being possible in the case of current coils though it is not shown in the diagram in order to avoid undue complication. This arrangement is possible since the vectoral sum of the pressures of any two phases is equal to the pressure of the third phase; the pressure coils (1) and (2) are fed direct from their respective transformers but the pressure coil (3) is fed by applying to it the resultant of the other two pressures.

Determination of Wave Form.

The determination of wave form has always been regarded as a subject of importance and in recent years it has attracted even greater attention since it has been realised that in the knowledge of exact wave form lies the solution of certain difficulties which have been met with from time to time. The problem is a difficult one owing to the rapidity with which successive waves follow each other, that is, owing to their small periodic time. The methods in use can conveniently be divided into two classes according as to whether an attempt is made to map out an individual wave or whether a series of observations are made on successive cycles; the latter method is only permissible on the assumption that successive waves are exactly similar, which is of course the case in many circumstances.

If an electrostatic voltmeter is connected continuously on to an alternating pressure it will give a reading showing the R.M.S. value of the pressure concerned; if, however, it is not connected to the circuit permanently but only for a brief instant each cycle, the reading will again be steady, assuming that the voltmeter is not leaky, if the contact is made at precisely the corresponding part of each succeeding cycle, and it will indicate the instantaneous value of the pressure at the particular part of the cycle concerned.

---

* If the power factors are unbalanced the statement is not exactly correct, since one axis may oscillate relatively to the other.

Thus if contact is made in each succeeding cycle at the instant
that the pressure is passing through its zero value the voltmeter
will read zero, and if at the instant it is passing through its
maximum value the voltmeter will show this maximum value
(*i.e.* a higher reading than would be observed if the voltmeter
were connected permanently across the pressure). A very convenient form of intermittent contact maker for the above purpose
can be made by mounting on the shaft of the alternator a brass
disc on one half of which a brush presses continuously, the other
half being provided with a projection which makes contact with
a second brush, insulated from the first, once in each revolution
(see Fig. 83). The second brush is carried on an arm, arranged

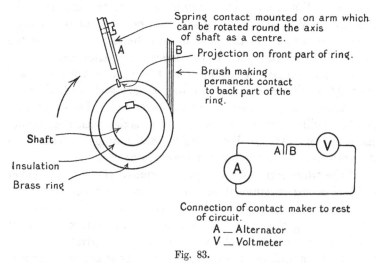

Fig. 83.

so as to be parallel with the shaft of the alternator, and which
can be rotated round the centre of the shaft as an axis, thus allowing
the intermittent contact to be made at any predetermined portion
of each cycle. The brush which makes contact with the projection
once in each revolution is composed of a piece of springy material
and after each contact it will have a tendency to vibrate; if this
vibration was allowed to continue it might result in the next
contact being made either a little too soon or too late and to
prevent this occurring the spring passes, for a portion of its path,
parallel to a more substantial brass plate which very effectively

damps out any vibration of the spring before the time for the next
contact arrives. A fixed brass disc is necessary so that the position
of the movable brush can be read off and this should be graduated
so that 360 electrical degrees correspond to the angular distance
between two north poles of the alternator field system. In many
cases the alternator giving the supply is not accessible and the
contact maker can then be fastened on to the shaft of a small
synchronous motor driven off the supply whose wave form it is
desired to investigate. Care must be taken that the synchronous
motor does not hunt. It is not essential that an electrostatic
voltmeter be used in the above test, a permanent magnet moving
coil instrument will do quite well, though the pointer is likely to
be in a continual state of vibration owing to the fact that the
deflection is due to a series of impulses rather than to a steady
force; the trouble on this account will be less the greater the
inertia of the moving parts of the indicating instrument.

It is also important to note that this type of voltmeter will
not indicate the true pressure on its ordinary scale and will need
to be calibrated for the circumstances of the test. The calibration
can be readily effected by connecting in series with the contact
maker a known steady direct pressure and noting the reading on
the voltmeter while the contact maker is running.

If the wave form of a current is required the contact maker
and indicating instrument should be connected across a non-
inductive shunt through which the current is passing, the result
obtained being the wave form of the pressure across the shunt
which will be similar to the wave form of the current. Such a
method of determining a wave form is laborious and great care is
necessary to obtain good results but a very convenient modification
of the method has been made use of by M. Hospitalier in the
instrument which he terms the ondograph; in this instrument
a recording voltmeter is made use of and the contact is moved
round automatically in order to obtain the pressures at different
parts of the cycle (instead of by hand as is necessary in the simple
apparatus described above)*.

The apparatus consists of a small synchronous motor driving
a rotating contact maker through a gearing such that (for example)

* See *J. I. E. E.*, Vol. 33, p. 75.

while the motor makes 1000 revolutions the contact maker only
makes 999 revolutions. If the position of the brush pressing on
the contact is stationary the actual contact will occur a little
later in each succeeding cycle. The contact used is not of the type
described above but is arranged as shown in Fig. 84 in which a
developed view of the circular rotating portion is given; it will
be seen that a condenser is for a brief instant connected across
the pressure to be measured and is then connected to the voltmeter
for the rest of the revolution of the contact maker. In the case
of the gearing ratio given above the contact maker will make

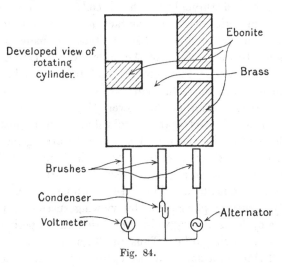

Fig. 84.

contact at $\frac{1}{1000}$ part of a cycle later in each successive cycle and
thus will have made contact at every phase of a cycle when 1000
cycles have elapsed, that is in 20 seconds if the motor has two
poles and the frequency of the pressure is 50 cycles per second.
The needle of the recording instrument will therefore pass through
a series of deflections which, if plotted against a time scale, will
be an exact reproduction of the wave form of the pressure under
test; the time scale is obtained by arranging that the paper on
which the pen of the recorder marks is moved along at a uniform
rate, in a direction at right angles to the line of motion of the pen,
by means of the motor. Readers should realise that the instrument does not follow one cycle but really takes a little bit out of

1000 successive cycles and this can only be done when the successive cycles are practically similar. If a current wave is required the instrument is used in a very similar way and the pressure wave taken across a non-inductive shunt through which the current is flowing.

Working on similar lines a puissancegraphe, or power wave recorder, may be constructed and this instrument is a particularly useful adaptation of the principle of testing under notice. To obtain the power wave the permanent magnet recording instrument is replaced by one of the dynamometer type and the main current is passed through the fixed coil; the moving coil is connected, in series with the contact maker and a resistance, across the mains so that when contact is made a current proportional to the pressure of the circuit passes through it, the resulting deflection is then proportional to the product of the instantaneous values of the current and pressure, that is to the instantaneous value of the power. The ondograph and the puissancegraphe are comparatively robust and portable instruments which do not require the use of any optical arrangements or of a dark room and their operation requires little more skill than is necessary for the use of any other recording instrument.

The principle made use of in these instruments of investigating a rapidly recurring phenomena by viewing successive phases in successive cycles is of very wide application and is perhaps not so well known as it should be. Thus an alternating current arc may be viewed through a hole in a disc carried by an induction motor run off the same circuit, the small slip of the motor ensuring that each cycle of operations is viewed at a slightly later phase than the previous one, the nett result being that the springing across, dying away and respringing across of the alternate current arc apparently takes place quite slowly.

Another modification of the same idea consists of mounting on the shaft of an alternate current motor, a disc having alternate black and white sectors, the number of each being equal to the number of poles of the motor, and illuminating it by an arc lamp operated by current obtained from the same source of supply as is being used to drive the motor.

The arc gives periods of maximum illumination twice every

cycle and, if the motor is rotating synchronously with the supply, the black and white sectors will appear to be stationary (though somewhat blurred); the explanation is, that in the time which elapses between one period of maximum illumination and the next, one black sector has moved into the position previously occupied by the one in front. If the motor is of the induction type and, therefore, rotates at less than synchronous speed, the sectors will appear to rotate slowly backwards and from the apparent rate of rotation the slip may be ascertained. Hunting in synchronous machines and the effect of variation of load can be visually studied in this way.

Oscillographs. If we attempt to determine a wave form by making observations of an individual wave the problem is more difficult than in the last case owing to the rapidity with which the waves follow each other. It is necessary to have an indicating instrument (*i.e.* a galvanometer) the inertia of whose moving parts is very low, otherwise it will be useless to attempt to follow the small rapid variations which occur during a wave; in addition, the moving parts must be very well damped or ripples will be observed which are due to the natural oscillation of the galvanometer and are not present in the actual wave under examination.

Perhaps the instrument which is best known in this country, and which is based on the principle of following the individual wave, is the oscillograph invented by Mr Duddell and manufactured by the Cambridge Scientific Instrument Co., Ltd.

Views of the galvanometer (or vibrator as it is usually termed) which is used in this instrument are given in Fig. 85, from which it will be seen that the current, whose wave form it is desired to observe, is led into a fine phosphor bronze strip P which passes over an ivory guide B and then through the narrow gap between the pole piece I and the iron separator S; from thence the strip passes over another ivory guide piece C, round the ivory pulley D, finally returning by a similar path between the other pole piece and the separator S. The strips are kept tight and clear of the pole pieces and separator since they are firmly secured at the top and the ivory pulley is pulled vertically downwards by the spring shown.

The path of the magnetic lines through the pole pieces and gap is shown by the dotted line MM and when current is passed through the strip, since it must flow upwards in one side and downwards in the other, one side of the strip will move slightly forwards from the brass support B and the other side of the strip will move slightly backwards towards the brass support. A small mirror Z is secured to the strips and this will be tilted one way or the other around a vertical axis when current passes through the strips. We thus have a galvanometer of high natural period (about $\frac{1}{10000}$ part of a second in the high frequency type of instrument) and small inertia, and if the movement is effectively damped by immersion in oil of suitable viscosity we have a device capable of following wave forms in an accurate manner. In practice the angular deflection of the mirror is always kept very small thus securing approximate proportionality between the current flowing through the strip and the resulting deflection. The magnet used to produce the field may either be of the permanent or electro type, the latter is usually preferred on account of the greater strength of field obtainable, but, for high tension purposes, the permanent magnet is used because of the facilities it offers for insulation. In recent patterns of the instrument two distinct movements are fitted thus allowing of the simultaneous observation of current and pressure curves; the oil bath has also been considerably enlarged in size, and each of the vibrators can readily be removed from the bath for examination and repair, a great advantage compared with the older type. In addition to the two moving mirrors, one on each movement, a third stationary mirror is also fitted so as to give a zero line.

The motions of the spot of light reflected from the mirror of the oscillograph may be observed either visually or photographically and the latter method will be considered first.

When the mirror tilts round a vertical axis the spot of reflected light will move to and fro very rapidly in a horizontal line; owing, however, to the persistence of vision, it will be impossible to observe the varying deflection when the instrument is connected to circuits of ordinary frequency, a line of light alone being observed. It is necessary therefore to find a method of impressing a time axis at right angles to the direction of oscillation of the spot of light and this may be effected by allowing the spot to fall on a strip of rapid photographic paper or film which is moving vertically at a uniform rate, due care being taken to prevent the action of extraneous light on the sensitive paper or film. A special apparatus

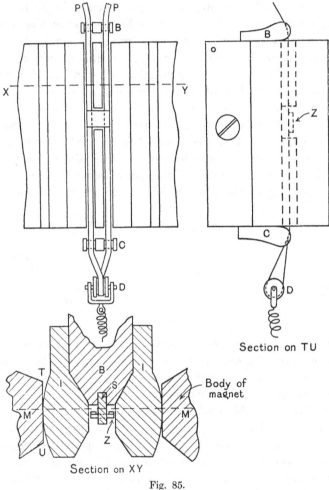

Section on TU

Section on XY

Fig. 85.

may be obtained for this purpose in which a film similar to that used for cinematographic purposes is made use of, but quite useful results may be obtained by using a light tight box inside which is a drum (preferably faced with linoleum for the easy insertion of drawing pins) which can be driven at a uniform rate by an external motor. Pieces of rapid bromide paper may be pinned to the drum and the oscillating spot of light is allowed to impinge on the surface of the paper through a narrow slit in the front of the case containing the drum. A falling photographic plate may also be made use of to record the waves. This photographic method of observation is not only available when successive waves are similar but also when they differ from each other, as, for instance, when observations are made of the effects of switching load on and off a cable.

The visual method of observation, which can only be used when successive waves are practically similar, is based upon the persistence of vision. We have already seen that the spot of light oscillates rapidly in a horizontal line, and if the spot is caused to fall on a second mirror which can be moved about a horizontal axis, see Fig. 86, the position of the spot reflected from this mirror will depend upon the angle which the mirror makes with the

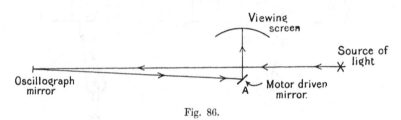

Fig. 86.

incident ray. If the second mirror is for the moment regarded as being at rest the spot of light will be reflected on to the viewing screen and will there oscillate in a line at right angles to the plane of the diagram; if the mirror is now turned about its axis the spot will be given an additional motion in the plane of the diagram, that is we shall have impressed on the spot a time scale. In order to obtain the true shape of the wave it will be necessary that the motion of the spot produced by the mirror A shall be uniform during the duration of the wave. In practice the mirror is caused to oscillate by a cam which is driven by a small synchronous motor making one revolution per two cycles of the applied pressure. The cam is shaped so that for about one and a half cycles the mirror is tilted in one direction at a uniform rate and is then quickly tilted back to its initial position during the remaining half cycle, these operations being repeated as long as the motor continues to run. During the period in which the mirror is making its quick return the spot of light is cut off by a revolving diaphragm. If the test is being made on a 50 cycle circuit we shall have the spot of light flashing across the screen and tracing out the wave 25 times per second and, owing to the persistence of vision, the eye will receive a steady impression of the wave.

The complete connections of the oscillograph to the external circuit are shown in Fig. 87 and it is worthy of note that in connecting up the two sides of the instrument, for the simultaneous observation of pressure and current waves, care must be taken to arrange that the pressure between the two sides is as small as possible, this precaution being necessary owing to the small clearances between the strips and the poles of the magnet. Other oscillographs worthy of attention are Irwin's hot wire instrument and the cathode ray oscillograph*.

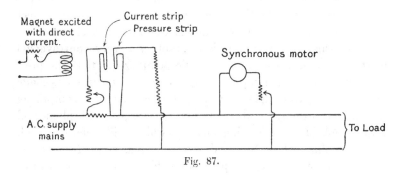

Fig. 87.

Supply Meters for Alternating Current Circuits.

On direct current circuits meters intended for the measurement of energy should take into account three factors: current, pressure and time, and those meters which directly take into account all three factors are known as watt hour or energy meters. In most cases, however, the supply pressure is constant within a few per cent. and some meters are arranged to work on the assumption

* For the former see *J. I. E. E.*, Vol. 39, p. 617, and for the latter the *Electrician*, Vol. 72, p. 171.

of constant pressure; they take into direct account only current and time and are known as coulomb or quantity meters. It will be evident however that, on the assumption of constant pressure, the dials can be calibrated directly in Board of Trade Units.

In alternating current circuits the measurement of energy is rather more complicated, because, in addition to the factors mentioned above, the relative phase of pressure and current must be taken into account and, since this quantity is very variable in different circuits, and also in the same circuit from time to time, the energy consumed is very far from being co-jointly proportional to current, pressure and time. This fact renders a coulomb type meter practically useless and so we find meters which are provided with a pressure coil almost invariably used. The pressure coil is perhaps best regarded as being necessary in order that the phase difference between pressure and current may be taken into account, but, since it is present, it is desirable to use it also for the purpose of taking pressure into account, thus making a watt hour type of meter. Such meters only take into account the working component of the current, the wattless component does not have any effect on the reading of the meter and is therefore not charged for. Whether this state of affairs is desirable is a very controversial matter but it is perhaps worth while pointing out that the price charged per unit by a supply company may be regarded as made up of two parts: (1) the preparation charges, comprising the annual interest and depreciation, etc. on the plant and mains, and (2) the running charges which will comprise such things as the cost of fuel and stores, wages, etc. Now if the consumer takes energy at a poor power factor he certainly causes an increase in the first item since more expensive generators must be installed and larger mains laid down, and these considerations point to the desirability of a consumer being charged not only for the energy consumed but also for the wattless current taken*.

Alternating current supply meters of the coulomb type could, if desired, be used with sufficient accuracy on certain purely lighting loads.

The types of watt hour supply meters most commonly used on alternating current circuits are:

* See *J. I. E. E.*, Vol. 51, p. 270.

(1) Motor meters of the induction type.

(2) Motor meters of the Thomson type.

(3) Clock meters of the Aron type.

Types (2) and (3) are essentially the same as are in use on direct current circuits and since they are well known their description is omitted from the present work. It is, however, desirable to point out the necessity for arranging that the current in the pressure circuit and the current in the current circuit have no phase difference when the load circuit is non-inductive and this is brought about by making the pressure circuit as non-inductive as possible. Since meters of the induction type are only applicable for use on alternating current circuits, and in such cases have a very extensive application, it will be well to briefly consider their operation. When two solenoids are arranged with their axes at a suitable angle (90° in a two pole case) they will, if traversed by two phase alternating currents, produce a rotating magnetic field. This rotating magnetic field may be used to induce current in a pivoted metal disc, and the interaction between the rotating field and the induced currents in the disc will cause a torque tending to turn the disc in the same direction as the field is rotating. The production of torque in this way is fully dealt with in Chapter X, where the induction motor is considered, and the induction type meter is simply a specialised form of the induction motor. Consider a laminated carcase, having a profile of the shape shown in Fig. 88, and let the two opposite poles C and C_1 be wound with a few turns of stout wire placed in series with the main circuit, and the two remaining poles V and V_1 be wound with many turns of fine wire connected across the mains. Neglecting the minor considerations of losses due to resistance and to the core, the magnetic field produced by the coils C and C_1 will be in phase with the current and, if the load circuit is for the moment regarded as being non-inductive, in phase with the pressure of the circuit. The field produced by the pressure coils V and V_1 will be 90° behind the pressure as regards phase, and thus we have the conditions laid down above as being necessary for the production of a rotating field completely fulfilled. The simple meter can be completed by the addition of a suitable disc, placed so as to be cut by the rotating field, mounted on a spindle

which is free to rotate and provided with a suitable counting
mechanism. In practice an eddy current damping device will
also be necessary in order to prevent the meter from running too
fast. The exact theory of the induction meter is rather complicated
and it is unnecessary to enter into it at this stage, it will be
sufficient to point out that both theory and practice agree in

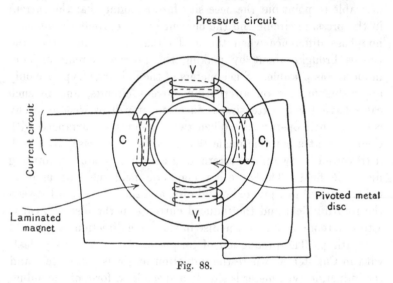

Fig. 88.

indicating that the speed of revolution of a disc working under
the above conditions is very approximately proportional to the

product of the strengths of the
component fields multiplied by
the sine of the angle of phase
difference between them. The
strengths of the two component
fields are proportional to the
current and pressure respectively,
and the sine of the angle of

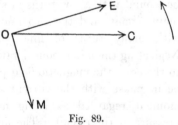

Fig. 89.

phase difference between the fields will be equal to the cosine of
the angle of phase difference between pressure and current in the
external circuit*. The speed of revolution of the disc will therefore

* In Fig. 89 OC represents the phase of the main current and also the phase of
the magnetic field which it produces, OE represents the phase of the pressure and

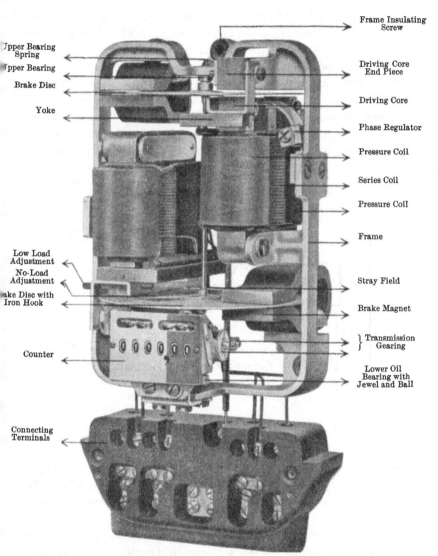

Fig. 90.   Two phase or Three phase Three-wire Meter.   Diagrammatic Sketch.
(Suitable for unbalanced loads, neutral not available.)

OM the phase of the field it produces. The angle COE is the phase difference between pressure and current in the external circuit and the angle COM is the angle between the two fluxes; it will be readily seen that $\sin COM$ is equal to $\cos COE$.

be proportional to the power passing through the meter, and the number of revolutions made in a certain time will be proportional to the energy taken by the main circuit in that time.

One of the chief difficulties met with in the commercial application of the induction type meter is in obtaining the two components of the magnetic field with a phase difference of 90° when the load is non-inductive. This difficulty chiefly arises from the resistance of the pressure coil and special devices are almost invariably used to prevent an inaccuracy, which is most noticeable on loads of low power factor, due to this cause. Care must also be taken that the meter does not run on no load when the shunt coil

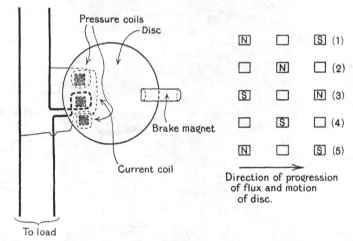

Fig. 91.

alone is excited. A modern example of the induction type supply meter as made by the Electrical Company, Ltd., to whom the author is indebted for permission to use the illustration, is shown in Fig. 90. The meter is intended for use on either two or three phase unbalanced circuits, and consists of two separate movements driving on to a common spindle. The electrical connections of one portion of the meter are shown in Fig. 91 from which it will be noticed that the two cores which are wound with the pressure coils are arranged so that at any instant opposite polarities are produced at the top ends.

The polarities of the top ends of the three cores at successive instants, differing by a quarter of a cycle, during one cycle are indicated in Fig. 91 (which is drawn for non-inductive load), and the gradual progression of flux from left to right is clearly seen. This motion of the flux will give rise, as in the simple case already considered, to induced currents in the disc thus producing the required torque. As an alternative arrangement to that shown in the diagram one pressure and two current coils may be utilised, the latter being on the outer limbs; this arrangement is also used by the Electrical Company.

EXAMPLES

1. If the circuit of a voltmeter of the moving iron type has a resistance of 2000 ohms and an inductance of ·5 henry, determine the percentage error when used on alternating current circuits at 25 and 50 cycles per second respectively, if it is assumed to read correctly on a direct current circuit. Assume that there is no other inaccuracy than that due to the varying impedance of the coil. *Answer.* ·075 % low; ·3 % low.

2. If the phase error of a dynamometer type wattmeter is ·2°, determine the percentage error introduced in the reading when the instrument is working on an inductive circuit whose power factor is ·1. *Answer.* 3·5 % high.

3. How would you make use of a low reading ammeter to indicate frequency on a constant pressure alternating current circuit? Mention any auxiliary appliances needed, give a diagram of connections and make a rough sketch of the scale of the instrument when used as a frequency meter.

4. Make a sketch of a drum and containing box for use in photographically recording oscillographic results as indicated on page 133. What distance on the time scale would represent one cycle of a 50 cycle circuit if the drum is 6 inches in diameter and is revolving at 80 revolutions per minute?

Answer. ·5″.

CHAPTER VII

ALTERNATORS

For the purpose of generating alternating currents three types of machines have been extensively used, these are as follows:

(1) Alternators of the moving armature type,

(2) Alternators of the inductor type,

(3) Alternators of the moving field type.

Alternators of the moving armature type are those in which the general arrangements are similar to the usual arrangement of direct current generators with the exception that the commutator is replaced by two or more slip rings, the exact number of rings depending upon the number of phases for which the machine is wound. The alternating current is collected from the slip rings by means of brushes, and this type is still frequently used for small and moderate sized machines working at low pressure.

When machines are required for high pressure work or for the generation of considerable currents, difficulties arise with the moving armature type of machine. In the first place the convenient and satisfactory collection of current becomes a difficult matter, and this can be avoided by the use of a machine in which the armature winding is stationary. Again, with high pressure machines, the difficulty of proper insulation of the armature conductors rapidly increases with rise of pressure and here also the advantages of the stationary armature are obvious. When machines are required for the generation of both alternating and direct currents, and also in machines of the rotary converter type, a moving armature type of machine is of necessity made use of because of considerations arising from the direct current side.

In order to minimise the difficulties mentioned above the inductor type of machine has been introduced. This is a type of machine in which there are no moving windings, the only parts moving (apart from the shaft, etc.) being masses of laminated iron known as inductors. The general arrangements of such a machine are indicated in Fig. 92, from which it is seen that the machine

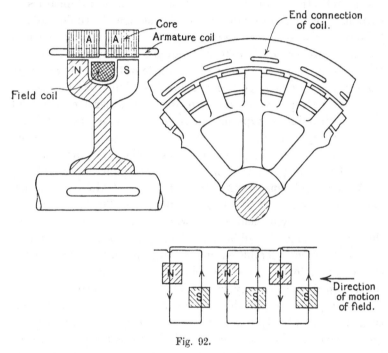

Fig. 92.

comprises a laminated stationary armature core A containing an armature winding which, in the diagram, is shown of a type suitable for single phase working: the field magnet consists of a double crown of poles, preferably provided with laminated pole tips, the excitation being provided by a single large stationary field coil whose plane is at right angles to the shaft. This field winding is supported from the armature frame and of course does not touch the revolving part at any point. The crown of poles on one side of the field coil are all of north polarity and those on the other side of the field coil are all of south polarity, further, the poles on one side are staggered relatively to those on the other. The reason for this staggering of the poles will be seen if an examination is made of the developed view shown at the bottom of Fig. 92; if poles of opposite polarity were opposite to each other, opposing pressures would be developed in the same conductor, but the arrangement actually adopted ensures that all the pressures assist each other. In this type of machine all the difficulties of current collection are removed and at first sight it would seem to be ideal, but in practice certain objections to the type have arisen and have prevented its extensive adoption. Amongst these are the following: (1) on account of the whole of the exciting ampere turns being in one large spool, a break or burn out in the coil can only be repaired at the expense of considerable time and trouble, (2) inductor type machines are very large and heavy for their output, and finally, (3) the regulating properties of such machines are, as a rule, poor and this, while excellent from the point of view of parallel running, is on the whole undesirable.

Moving field type alternators. The type of alternator which has become standard for large outputs and for high pressure working is the moving field type, in which the armature is stationary thus overcoming the difficulty of current collection, and also lessening to a considerable extent the difficulty of proper insulation of the conductors. Since the field coil is moving it is necessary to use slip rings to convey to the field the direct current used for excitation purposes, but this current is not of great amount and is at a comparatively low pressure and so does not involve any great difficulty. Diagrams of a machine of this type arranged for low speed work (that is having a peripheral speed up to say 5000 feet per minute) are shown in Fig. 93, the material for which has been supplied by the Phoenix Dynamo Manufacturing Co., Ltd.

The stator housing H carries the laminated core C which has the usual series of slots on the inner surface for the reception of the armature winding. The winding is of the three phase type with two shapes of end connections, one straight out and the other bent up, the ends of the windings coming out to terminals insulated with porcelain bushes as indicated at E.

Turning now to the magnet wheel, the steel poles P are bolted to the steel flywheel-like rim R which in turn is keyed to the shaft.

Laminated pole shoes S are secured to the poles and these shoes, in addition to providing a proper distribution of the flux, also secure the field coils in position. The field coils are composed of

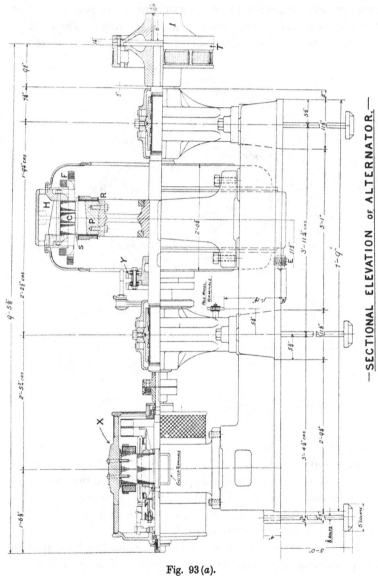

Fig. 93 (a).

copper strip wound on edge, a type of construction which is well adapted to resist effectively the large centrifugal forces which are called into play and, at the same time, allow of the ready dissipation of the heat generated within the coils. The field coils (20 in number) are connected in series and current is passed through them by means of brushes bearing on the slip rings. The machine illustrated is provided with a direct coupled exciter X seated on

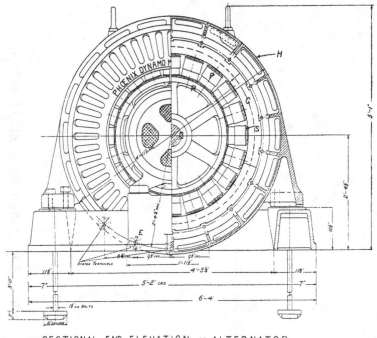

— SECTIONAL END ELEVATION of ALTERNATOR.—

Fig. 93 (b).

an extension of the bed plate, and the field current of the alternator is very conveniently varied by varying the field current of the exciter; this is readily and cheaply effected since the current to be controlled is small. The alternator is arranged for direct coupling to an engine and for this purpose is provided with a flexible and insulating coupling at I.

In recent years very considerable attention has been paid to the production of turbine driven alternators of large power, and in connection with such machines designers have found many difficult problems to solve. In respect to the general arrangement, high speed alternators are constructed on similar lines to low speed machines, but the two classes differ considerably in details.

Dealing first with the rotating field systems of high speed machines, two distinct types of field magnets are employed:

(1)   The salient pole type, and

(2)   The smooth cylindrical core type.

The salient pole type.   On account of the very high speed of revolution found in steam turbines, it is necessary to keep down the diameter of the rotor as much as possible in order to obtain reasonable peripheral speeds, and it is then necessary to use a long

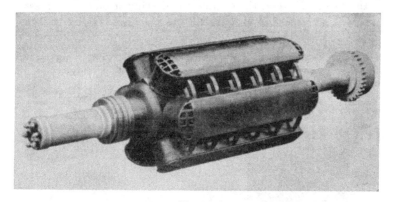

Fig. 94.

core in order to obtain the required output; consequently the ratio of core diameter to core length is considerably less in turbo-machines than in slow speed alternators. Even under these conditions the centrifugal forces called into play are very considerable and so a strongly braced construction is necessary. Fig. 94 represents a salient pole type of field magnet forming part of a turbine driven alternator, designed to give an output of 8500 K.V.A. at 50 cycles per second and 11,400 volts when running at 1000 revolutions per minute. For the photograph used in the preparation of this figure, and also for those used in the preparation of Figs. 96 and 98, the author is indebted to Messrs Dick, Kerr and Co. of Preston, who have also supplied much information concerning the machines illustrated. The main body of the rotor consists of steel cast under pressure and with projecting poles; this is bored out to take the shaft and is secured in position by two end pieces, also of cast steel, which are forced on to the shaft and are firmly secured to the central portion. The pole shoe is laminated, the laminations being shaped so as to dovetail into the body of the pole and also to hold the sides of the field coil in position. The ends of the field coil are kept down by end pieces of cast steel dovetailing into the end pieces of the body of the core already mentioned. The field coils are also braced together by the phosphor bronze brackets which are clearly seen in the figure and which are dovetailed into the rotor body. The rotor winding consists of copper strip, wound on edge, and insulated with mica and paper, the complete coil being strongly compressed before being placed in position on the machine.

Fig. 95.

The smooth cylindrical core type. Fig. 95 represents a field magnet of this type as made by the British Westinghouse Electric and Manufacturing Co., Ltd., to whom the author is indebted for the photograph and for information concerning the same.

In this case the field core is built of sheet steel stampings bolted up between end plates shrunk on to the steel shaft. It is claimed by the makers that this type of field is quiet in operation and wastes the minimum amount of energy in air friction. The field coils are of copper strip placed in slots, the sides of the coils being held down by the overhanging tips of the teeth and by metal wedges, and the ends by special binding bands.

The rotor shown in Fig. 96 is also of the smooth core type, but the method of construction is very interesting and entirely different from the one last described. In this case the rotor and shaft are formed out of one solid steel forging, which is accurately machined all over. The sides of the coils are placed in slots machined in the body of the core and are held in position by wedges of manganese bronze. The end portions of the field coils are housed in bells of manganese bronze which are forced on to the shaft.

Fig. 96.

Stators of turbo-alternators. In connection with the stators of high speed machines the chief difficulty has been experienced with the mechanical support of the end connections. On account of the high peripheral speed these are much longer than in slow speed machines, and this makes them very susceptible to damage in the event of the machine being short circuited when fully

excited, or if it is paralleled out of proper phase relationship with
the bus bar pressure*.

The difficulty has been successfully surmounted by providing
a very complete system of clamping supports such as are shown
in Fig. 96 (Messrs Dick, Kerr and Co.) and Fig. 97 (The British
Westinghouse Co.).

Fig. 97.

* It has frequently been stated that the short circuit current of an alternator
is but three or four times the normal full load current, and this is true if the machine
is run up to speed and the excitation gradually put on while the armature winding
is short circuited, the comparatively small value of the current in the armature
under these conditions being largely due to the demagnetising of the field by the
lagging armature current. Experience has shown, however, that if the short
circuit comes on while the machine is fully excited, much larger currents flow
through the armature for a short time; this is owing to the field magnets taking an
appreciable time to become demagnetised, and it is because the machine may have
to face such conditions that the difficulty with the end connections arises. See
a paper by Prof. Miles Walker, *J. I. E. E.*, Vol. 45, p. 295.

In large machines, such as those illustrated, the dissipation of
the heat generated is a very serious problem which can only be
solved by a very efficient system of ventilation. In general, cool
air is drawn (either by internal or external fans) into the machine
from outside the engine room (in manufacturing towns it is
necessary to filter the air) and is then systematically forced through

Fig. 98.

the rotor and stator via a carefully thought-out system of ventilating ducts. Ducts must be provided both in lines parallel to the
shaft and in planes at right angles to the shaft. The warm air is
usually discharged into the spaces of the box type stator housing,
which are arranged at the back of the stampings, and from there
either into the engine room or directly to the outside of the
building. The fans, the ventilating ducts in the armature which
are at right angles to the shaft, and the spaces in the housing are
well shown in Fig. 98, which represents a portion of a large machine

which has its stator divisible into four sections for convenience in shipment.

Armature windings for alternators. The armature windings met with in alternators are of many types and in an elementary book it is not possible to go deeply into the subject. In modern machines the windings are almost invariably placed in slots uniformly distributed round the armature core, the number of slots per pole commonly varying from six to twelve. Consider the case of a core provided with six slots per pole and used for a single phase winding; in all probability the whole of the slots would not be utilised for the winding, some being left idle to avoid what has been termed differential action. In the case under consideration probably four slots per pole would be used. If a similar core was made use of for a two phase winding the whole of the slots would be used, that is three slots per pole per phase, and, as we pass round the armature, we should successively encounter three slots occupied by the wires of one phase and then three slots occupied by the wires of the second phase. Again, for a three phase winding, the whole of the slots would be utilised and this time we should have two slots per pole per phase, and, as we pass round the armature, we should encounter successively two slots devoted to the first phase, then two slots devoted to the second phase and finally two slots devoted to the third phase and so on.

It is usual in alternators to connect the whole of the conductors in any one phase in series, thus providing but one path through each phase; this prevents (at any rate if the star method of interconnection is adopted) the formation of internal eddy currents which would be likely to occur if there were two or more paths per phase. All the conductors in one phase must therefore have their end connections arranged so that the induced pressures assist each other, and it is chiefly in the arrangement of these end connections that the different types of alternator windings arise. Consider first a machine wound so as to produce a high pressure (say several thousand volts); in this case the individual conductors are not likely to be of large cross-section but many conductors will be required in series and quite a number of conductors will be in each slot. In such circumstances it is usual to employ what may be termed a coil winding.

Thus, in Fig. 99, let A be a group of conductors situated (at some instant) under a north pole, and B an equal number of conductors situated under an adjacent pole of opposite polarity, a coil may be made by joining the various conductors in the manner indicated and this coil may be connected by a single wire to adjacent coils formed in a similar way by grouping together the conductors under other pairs of poles. It is not necessary that the whole of the conductors in group A be in one slot, as is perhaps indicated in the figure, they may be distributed in two or even more slots if the number of conductors warrants so doing. There is no difference in principle, one coil side is simply distributed over several slots instead of being concentrated in a single slot.

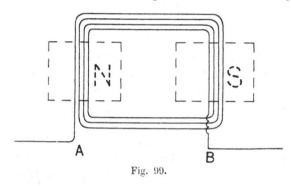

Fig. 99.

If the arrangement indicated above be adopted there will be comparatively little difference of pressure between the conductors in any one slot, since they are electrically adjacent, hence we find the individual conductors but lightly insulated from each other, say by an impregnated cotton lapping or braiding. Between the conductors in any one slot and the core there may exist considerable differences of pressure and the insulation between these two parts (known as the slot insulation) is made much heavier, the amount of insulation of course being greater the higher the pressure for which the machine is intended; it is perhaps best made of a jointless tube of micanite (see Fig. 100). If the system of end connections described above is adopted two distinct arrangements are possible: consider a single phase winding on a core with six slots per pole, four being actually used for the conductors. If the end connections are as in case A, Fig. 101, there will be as many coils as poles, each coil side being distributed over two slots; this arrangement has been termed a whole coil winding. On the other hand, if the end connections are arranged as in case *B*,

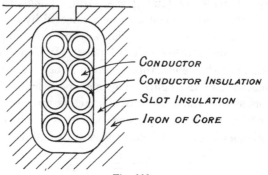

Fig. 100.

there will be half as many coils as poles and each coil side will be distributed over four slots; this has been termed a hemitropic winding. The pressures generated by the two windings will be

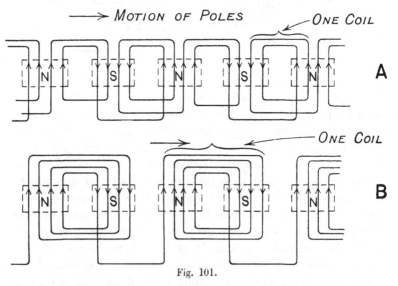

Fig. 101.

exactly equal but the average length of the end connections will be slightly greater in the latter case, hence more copper will be required and the winding will have a rather greater resistance.

The shape and arrangement of the end connections assume greater importance in two and three phase machines and, since the latter case is by far the most common, it will be well to consider it in some detail. Here we have three distinct sets of end connections and they will, at certain instants, be at very different potentials and so must be well insulated from each other; the insulation is effected by suitable wrappings and also by air spaces and to obtain these air spaces it is desirable to make use of at least two shapes of end connections.

Considering again the case of a core having six slots per pole, we shall now have two slots per pole per phase, the whole of the slots being utilised. If the end connections are arranged on the principle of the hemitropic winding mentioned above, it will be possible to keep the end connections of the three phases quite clear from each other if two shapes of end connections are used as indicated diagrammatically in Fig. 102. This is known as a two plane winding.

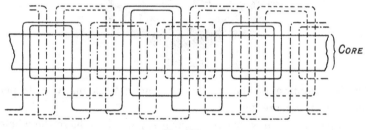

Fig. 102.

An actual photograph of such a winding is shown in Fig. 103 which has been kindly furnished by the Phoenix Dynamo Manufacturing Co. of Bradford. An examination will show that this illustration represents a winding having four slots per pole per phase and which is suitable for a twelve pole field. It represents the armature before the ends of the coils are taped up and the two types of coils are of the straight out and bent up patterns.

Notice particularly that when this type of winding is employed for three phase machines, alternate coils of one phase are of different shapes. Fig. 104 represents the result of using a whole coiled winding for the three phase case when the core has six slots per pole; it will be seen that three shapes of end connections are necessary, that is, a three plane winding must be used. In three phase machines therefore, simplicity in end connections is obtained

Fig. 103.

by the adoption of the winding having half as many coils per phase as there are poles.

In passing, it may be noted that it is possible to have the whole of the end connections of the same shape in a three phase machine,

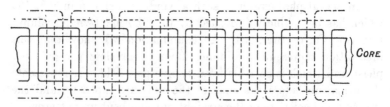

Fig. 104.   N.B.   Connections between successive coils of each phase are
not shown in order to avoid complication.

but the coil then becomes of a much more complicated shape and this system is not often made use of.

A three plane winding may, if desired, be used for the hemitropic type of end connection and we then arrive at the arrangement
shown diagrammatically in Fig. 105.   This winding permits of the
armature being split (say for convenience in transport) without
disturbing the coils, all that is necessary being the disconnection
of successive coils from each other and this can be very readily
carried out.

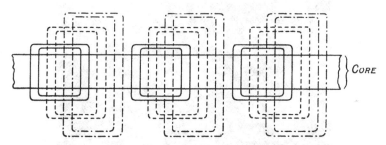

Fig. 105.   N.B.   Connections between successive coils of each phase are
not shown in order to avoid complication.

An actual winding carried out in this manner is shown in
Fig. 106 (the photograph for which has been supplied by the
Phoenix Dynamo Manufacturing Co., Ltd.), which represents a part
of the stator of a 600 K.W., 3000 volts, 50 cycle generator having
two slots per pole per phase and eight conductors per slot.

Such coil windings may be hand wound; in which case, in order
to secure a neat and systematic arrangement of the conductors in
the slot, the pin method of winding may be employed.   This
consists of securing the slot insulating tubes in the slots and then
placing within each tube rods to the number and diameter of the
conductors which are to go into the slot.   As each turn of the
conductor is pulled through the slot it displaces one of the rods,
the end connections at the same time being neatly arranged and
eventually taped up.

If the slot insulation is split, instead of being in the form of
a closed tube, it is possible to make use of former wound coils and
the interior of the coil can then be filled in solidly with a suitable
compound, a process that has been found to prolong the life of
the insulation in certain cases, possibly owing to the absence of
a brush discharge within the coil.

In certain cases former wound coils have been used in conjunction with seamless tubes of slot insulation; to effect this the coil is wound, the end connections at one end are cut and straightened out, the two coil sides passed through the tubes which have already been inserted in the slots, and finally the cut wires are rejoined and the end connections suitably insulated and taped up.

Fig. 106.

Turning now to the armature windings for low tension and large current machines; the conductors will be fewer in number but of larger cross-section and what may be described as a bar winding is conveniently made use of. One form of winding on the above lines, which has been used extensively in connection with low pressure alternators and for wound rotors of induction motors, may be described as a two layer wave winding and, in appearance, is very similar to the wave windings used in direct current machines.

The general appearance of a winding carried out on these lines is shown in Fig. 107 which has also been supplied to the author by the Phoenix Dynamo Manufacturing Co., Ltd.

Fig. 107.

The methods considered above for the connection of the conductors of alternator windings are taken as being most typical of modern practice, but it should be realised that the matter has been considered in a far from exhaustive manner.

One other point in connection with the armature windings of alternators is perhaps worthy of mention and that is the correct inter-connection of the various phases for the star and mesh arrangement respectively.

Considering the case of star connection first, imagine a machine working on non-inductive load and with the conductors of one phase, at the instant under consideration, directly under the centres of the poles. At this instant the current will be at its maximum value in this phase, and, if it is regarded as flowing towards the neutral point, the remaining phases must be connected so that the current in them is flowing away from the neutral point (this is the case since there can be no accumulation of electricity at the neutral point). The correct ends of the phases can readily be starred from the above considerations and if a further examination is made with the poles in a different position it will be found that the same connections are, of course, still correct.

In a similar manner the correct ends to connect in the case of the mesh arrangement can be found since we know that the resultant pressure round the mesh, due to the fundamental wave, must at each instant be zero.

The conductors of one phase should be considered as lying in the centres of the pole faces, that is in the position of maximum induced pressure, and this phase must be connected so that its pressure opposes the pressures developed at the same instant in the remaining phases in the local circuit round the mesh.

In addition to the specialised alternator windings just considered, readers may again be reminded that alternating current may be obtained from any ordinary direct current winding if slip rings are connected at suitable points.

Pressure developed in Alternator Windings.

If the useful flux per pole, the number of conductors in series per phase, and the frequency of the pressure produced are represented by the letters F, N and f respectively, then we have the following relations:

Lines cut per conductor per cycle $= 2F$, and lines cut per conductor per second $= 2Ff$.

The average pressure produced per conductor is therefore $\frac{2Ff}{10^8}$ volts, and the average pressure produced per phase is $\frac{2FfN}{10^8}$ volts, if all the pressures in conductors connected in series are assumed to be in phase with each other.

Now in practice we are not much concerned as to the magnitude of the average pressure, it is the R.M.S. value of the pressure that is required, but if the pressure wave is assumed to be sinusoidal the ratio between the R.M.S. and the average values will be 1·11 and the R.M.S. value of the pressure generated will therefore be $\dfrac{2 \cdot 22 FfN}{10^8}$ volts.

The constant introduced into the expression is known as the Kapp coefficient and only has the value given under the particular circumstances mentioned above. A distributed winding will have a rather lower constant, whose value will also be affected by the distribution of flux in the gap; if the gap flux wave follows a curve which is more peaked than a sinusoidal wave the tendency will be to increase the constant and *vice versâ*.

Pressure drop and regulation of Alternators.

When the load on an ordinary type of alternator is increased, the field current and speed being kept constant, the terminal pressure falls, often to a considerable extent. It is convenient to recognise three sources of this drop in pressure:

(1) **The effect of armature resistance.**

(2) **The effect of armature reactance,** that is to say the effect of the armature current setting up a magnetic field in the immediate vicinity of the conductor. This field causes the armature to have inductance, as in a choking coil, and it follows that a certain amount of pressure is necessary to drive the current through the internal reactance occasioned by this inductance.

(3) **The effect of armature reaction.** This is also due to the magnetic effect of the armature current and arises from the fact that the current not only exerts an effect in the immediate vicinity of the conductor but also over a much wider region, namely, upon the field magnets themselves, thus modifying their strength and field distribution.

Effect of armature resistance in producing pressure drop. The drop produced in the pressure of an alternator owing to resistance is generally of small amount and can readily be calculated for any load if the resistance per phase of the winding, duly corrected for skin and temperature effects, is known. The resistance drop is always in phase with the armature current.

The effect of armature reactance in producing pressure drop.
As stated above the armature reactance is conveniently looked
upon as being produced by the armature current setting up a local
magnetic field. The numerical value of the inductance of an
armature winding can readily be calculated by using the method
due to Hobart which is given on page 35.

Having determined the inductance, the resulting pressure drop
for any load can be calculated by the usual formula $E = I\omega L$,
and the pressure drop due to the reactance will invariably be 90°
in front of the current as regards phase. In most modern machines
the numerical value of the reactance will be several times the
numerical value of the resistance.

If the speed and excitation of an alternator be kept constant
and the effect of armature reaction be, for the moment, neglected,
the generated pressure may be looked upon as constant at all
loads, but the terminal pressure will fall as the load increases due
to the increased pressure drops with increased current. The
terminal pressure at any load can be found by vectorally subtracting the two pressure drops mentioned above from the
generated pressure. Now, in practice, alternators work under
various load conditions, in some cases the load may be practically
non-inductive and the current is then in phase with the terminal
pressure, in other cases the load may be more or less inductive
and the current will then lag behind the terminal pressure. Therefore, while the two pressure drops always have the same phase
relationships to the current, they may have very different phase
relationships to the terminal pressure in different cases.

For any given value of the load current the two drops will
always have practically the same numerical values, yet, for different
values of the power factor in the load circuit, different terminal
pressures will result owing to the different angles at which the drops
will be subtracted from the generated pressure.

These points will be clearly seen on examination of the vector
diagrams shown in Fig. 108. In (a), which is drawn for a non-
inductive external load, the current is shown in phase with the
external pressure but of course it lags a little behind the generated
pressure on account of the inductance of the armature; OT
represents the terminal pressure and OC the current, the resistance

and reactance drops being represented respectively by OR and OX. On vectorally adding OX and OR to OT we get the generated pressure OG; an alternative way of looking at the matter is to think of OR and OX being subtracted from OG leaving the terminal pressure OT. Fig. 108 (b) shows the corresponding diagram for a purely inductive load, and Fig. 108 (c) the diagram for a partially inductive load. A consideration of the diagrams will show that the resistance drop is likely to have a greater effect on the pressure regulation on loads which are nearly or quite non-inductive, since under these conditions the resistance drop is nearly directly subtracted from the generated pressure, while on loads which are approximately purely inductive the resistance drop is nearly at right angles to the generated pressure and consequently will produce a less numerical result when subtracted.

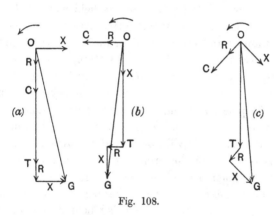

Fig. 108.

On the other hand, similar reasoning will indicate that the reactance drop is likely to be more important on loads which are very inductive. In practice, since the numerical value of the reactance drop is usually greater than the numerical value of the resistance drop, there is likely to be a greater difference between the generated and terminal pressures on inductive than on non-inductive loads, the value of the load current being taken as the same in each case.

Drop in pressure caused by armature reaction. In order to investigate the nature of armature reaction, which is due to the effect of the armature currents upon the poles of the field system, let us consider the case of a two pole machine of the moving armature type, the armature being arranged for the production of polyphase currents.

Assuming for the moment that the load is non-inductive, it will be realised that the conductors of the armature in which both pressure and current have considerable values are those passing in front of the poles at any instant. As the armature rotates successive conductors pass under the poles but the bands of current remain fixed in space and also remain of approximately constant total strength. The magneto-motive-force of these bands of current will tend to produce a magnetic field in the direction of the axis of the solenoid which we may regard as being formed by the bands. Thus in Fig. 109 (*a*), which is drawn for non-inductive load, the axis of the main field is shown by the continuous arrow, while the axis of the field which the armature tends to produce is shown by the dotted arrow. These directions are at right angles and we therefore deduce that armature reaction is likely to have but little effect on the strength of the main field, and consequently on the pressure generated, when the load is non-inductive. As a matter of fact the tendency is to distort the flux from the pole by strengthening one pole tip and weakening the other and this may, indirectly, cause a slight diminution of the total flux. Readers will notice that the armature reaction in this case is precisely of the same nature as that which occurs in direct current machines when the brushes are in the neutral position. Let us next consider what happens on a purely inductive load; in this case the currents will lag behind the pressures by 90°, the bands of maximum pressure will still occur under the pole faces but the bands of maximum current will not occur until the conductors have rotated 90° from the position corresponding to maximum pressure, that is the bands of maximum current will now occur in the spaces between the poles rather than under the poles. This state of affairs is indicated in Fig. 109 (*b*) in which the same notation is used for the arrows as before; it will be seen that the magneto-motive-force of the armature now directly opposes the main field, which will be considerably weakened on load, with consequent drop in the generated and terminal pressures, if the exciting current of the machine is

unaltered. Similar considerations will show that if the load takes
a leading current the result of armature reaction will be a
strengthening of the field with consequent rise of generated pressure.
In practice the load is likely to be to some extent, though not
completely, inductive, and in such cases we can imagine the total
current split up into working and idle components; it is the latter
that will be chiefly instrumental in weakening the field. For a
given value of the load current the demagnetisation of the field
will be proportional to the wattless component of the current,
that is to $\sin\phi$, where ϕ is the angle of phase difference between
generated pressure and the current. In modern work it is more
usual to have alternators of the moving field type, but even then
the same principles of armature reaction apply, though it must be

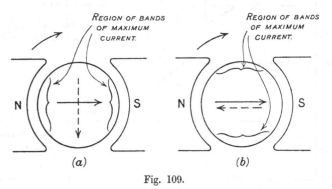

Fig. 109.

remembered that in such cases both the main field and the field
due to the armature current are revolving with the speed of the
field magnets instead of being stationary as in the case considered.

The ideas of the reader on the subject of pressure regulation of
alternators will perhaps be fixed by a consideration of the
characteristic curves shown in Fig. 110, which were obtained from
a three phase 10 k.w. machine running at constant speed and
excitation. In curve (a) the load was single phase and non-
inductive, the pressure drop with load being partly due to resistance
and partly due to reactance. Curve (b) was obtained with single
phase inductive load, the increased pressure drop in this case
being partly due to armature reaction, which lowered the generated
pressure, and partly due to the fact that the reactance drop is now

more directly subtracted from the generated pressure. In curve (c) the effect of a three phase inductive load is shown, the pressure being measured across the terminals of one phase as in the preceding cases; here we have greater armature reaction, since three currents are demagnetising the field in place of the single current in the previous case, the result being a greater fall in pressure. Note that the reactance and resistance drops will be the same for (c) as for (b) since the pressures are measured across the terminals of one phase in each case.

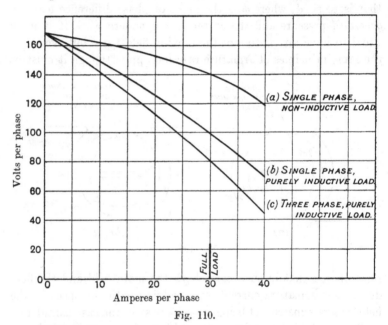

Fig. 110.

A numerical idea is sometimes attached to the term regulation and from this point of view the regulation of a machine may be defined as the value of the ratio

$$\frac{\text{Rise in pressure from full load to no load}}{\text{Full load terminal pressure}} \times 100\,\%,$$

the speed and excitation being kept constant throughout the test.

If an alternator is required to have good regulation it is desirable to arrange for the machine to have a strong and stiff field, that is a large flux per pole driven by plenty of field ampere turns, the latter condition being attained by having a comparatively long air gap. Since the flux per pole is large the armature turns will be kept down, thus minimising the reactance and resistance drops, and since the field ampere turns per pole are large the demagnetising ampere turns of the armature will have a comparatively small effect. The arrangement indicated would necessitate the use of a large and expensive field magnet and it should be realised that good inherent regulation can only be obtained by considerably increasing the cost of the machine. Professor Kloss has given the following values as being typical of the inherent regulations obtained in modern high speed machines*:

Power factor	Regulation
1·0	5 to 7 %
·9	12 to 15·5 %
·8	15 to 18 %

The effect of power factor on regulation is very clearly seen in the above figures; the increase of the number representing the regulation on the lower power factors is chiefly due to increased armature reaction but it is also, to some extent, due to the fact that the drop due to armature reactance becomes more and more nearly subtracted from the generated pressure as the power factor falls.

The power factor of the load which is put on an alternator influences very considerably the output which can be obtained from the machine at the prescribed pressure and without undue heating. In regard to the armature winding a small power factor means increased current and consequently increased heating for the same power; if we look upon the permissible temperature rise as being the same in each case, a lowering of the power factor will diminish the output if this is expressed in K.W., but will not affect it, from the point of view of heating, if it is expressed in K.V.A. Again, a lowering in the power factor means an increased pressure drop for reasons already given, and this tendency must be compensated for by increased ampere turns on the field; with low power factors, therefore, the load at which the field coils reach their maximum permissible temperature rise will be less than with high power factors. We may express these facts by saying that

* See *J. I. E. E.*, Vol. 42, p. 176.

a machine for a given output in K.W. will be larger, both in regard
to the armature and field, the lower the working power factor,
and will of course be correspondingly more expensive.

Automatic regulation of alternators. Owing to the poor inherent
regulation of alternators considerable difficulty has been experienced in maintaining a steady pressure on the bus bars of
central stations, particularly in the case of those stations in which
the load is very variable and is of an inductive nature. This
difficulty has led to many attempts to obtain a satisfactory system
of automatic regulation, either by the use of external regulators
acting on the field rheostats, or by some system of compounding.
Several manufacturers are now prepared to supply compounded
alternators and one of the most interesting systems is that used
by Messrs C. A. Parsons and Co., Ltd., to whom the author is
indebted for information concerning the same and also for Figs.
111 (a) and (b). The chief difficulty in the way of the production

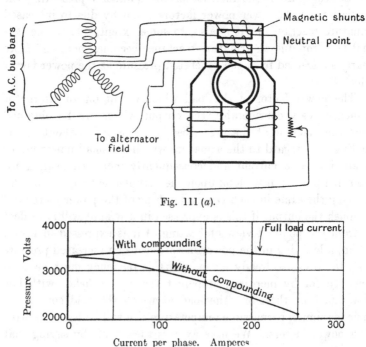

Fig. 111 (a).

Fig. 111 (b).

of a compounded alternator is due to the fact that the magnetic
effect of one half cycle of the alternating load current is, in general,
neutralised by the opposite magnetic effect during the next half
cycle. The firm mentioned has got over this difficulty by taking
advantage of the fact that if a mass of iron is already practically
magnetically saturated by a direct magnetising current and is then,
in addition, acted upon by an alternating current, the half cycle
of current which tends to still further magnetise the core can have
but little effect, while the half cycle which tends to demagnetise
the core can be very effective, the nett result being to diminish
the flux passing through the iron in question. To take advantage
of this principle the direct current exciter attached to the alternator
is provided with laminated magnetic shunts to the armature, as
indicated in the figure, and at no load much of the flux generated
by the field magnets of the exciter passes round this path instead
of passing through the exciter armature. As load increases on
the main machine the alternating load current passes round the
magnetic shunts thus diminishing the flux passing through them
and, indirectly, increasing the flux passing through the armature
of the exciter, thereby increasing the exciting current and pressure
of the alternator*.

That the arrangement is effective will be seen from the curves
given in Fig. 111 (b), which are load characteristic curves for a
three phase alternator, having an output of 1000 K.W. at 3300
volts, with, and without, the compounding device respectively,
the field rheostats being kept in a constant position through the
test.

Another method of compounding, which is due to Professor
Miles Walker and has been used by the British Westinghouse Co.,
Ltd., is well known and is described in the *Journal of the Institution
of Electrical Engineers*, Vol. 34, p. 431. It is of particular interest
since the armature reaction, usually a source of pressure drop, is
most ingeniously made use of to obtain the compounding effect.

* If the ampere turns of the exciter field are regarded as constant as load
increases (as a matter of fact they will increase somewhat), the effect of the
apparently increased reluctance of the shunt with increased load on the alternator
would be to diminish the total flux sent by the exciter while increasing the flux
sent through the exciter armature. The effect is similar to that which would
happen in an analogous electrical circuit.

This method is not available for use in cases when the power factor of the circuit is very low.

Finally, it may be stated that external regulators, actuated by the pressure or current of the line and working on the field rheostats, have also been used extensively for the automatic regulation of pressure.

The Tirrill regulator. One very interesting type of external regulator is the Tirrill regulator which is manufactured by the British Thomson-Houston Co., to whom the author is indebted for information concerning the same and for permission to use the diagrams shown in Fig. 112.

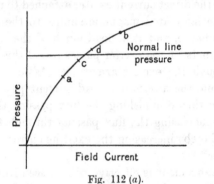

Fig. 112 (a).

Fig. 112 (b).

The essential features of the regulator comprise a resistance which is inserted in the field circuit or short circuited, as required, by means of a vibrating contact maker. The contact maker is operated by the line pressure and when the latter becomes too low the contact is closed thus increasing the field current, while when the line pressure becomes too high the contact is opened thus tending to lower the field current.

In Fig. 112 (*a*) let the graph represent the magnetisation curve of an alternator and imagine that when the field resistance is short circuited the pressure rises to the point *b* (if sufficient time is allowed), while when the resistance is inserted in the circuit the pressure falls to the point *a*. Suppose that the line pressure is too low, as at the point *c*, the short circuiting contacts will then be closed by the operation of the main control magnet, the resistance is cut out of the field circuit and the pressure begins to rise. If allowed to attain a steady value it will ultimately reach the point *b*, but long before this is attained the relay contacts will be opened again by the operation of the main control magnet, the pressure only attaining a point as *d*. As soon as the resistance is in the field the pressure will begin to fall only to commence to rise again when it has fallen to *c*. It is necessary to realise that the relay which is operating the field resistance is vibrating very rapidly with the result that the points *c* and *d* are very near together (the total variation of line pressure can be kept within 1 % under all conditions of load and power factor).

A complete diagram of connections of such a regulator is given in Fig. 112 (*b*), in which the regulator is shown acting on the field of the exciter instead of upon the field of the main machine. When the line pressure gets too low the plunger of the main control magnet falls thus closing the main contacts. This allows current to flow through the right-hand side of the winding of the relay, thereby neutralising the magnetisation of the latter, since the left-hand side of the relay winding is permanently excited so as to oppose the right-hand side. The arm of the relay then rises and short circuits the resistance in the exciter field and the exciter pressure and the main pressure both commence to rise. As these pressures rise above their normal value the exciter control magnet pulls its plunger down while the plunger of the main control magnet rises, thus giving a quick break on the main contacts; the right-hand winding of the relay magnet then becomes de-energised and its armature is pulled down by the left-hand winding thus reinserting the resistance in the exciter field (sparking at the relay contacts is prevented by means of the condenser shown). The whole of the above series of operations is rapidly repeated with the result that a very steady line pressure is maintained. The dotted winding on the main control magnet can be used, if desired, to compensate for line pressure drop.

Parallel Operation of Alternators.

In central stations it is customary to operate two or more alternators in parallel at times of heavy load and it is therefore desirable to investigate the behaviour of machines running under these conditions. Consider the case of two alternators (taken as single phase machines for the sake of simplicity) connected in parallel on to bus bars as indicated in Fig. 113; each of the machines will, in general, be supplying current to the external circuit, but there is, in addition, a local circuit composed of the armatures of the two machines (and indeed round any pair of machines if there are more than two on the bars) and it is the current which may flow in this local circuit which it will be desirable to investigate.

If the two machines are in absolutely the correct phase for parallel running their pressures will be in exact opposition as regards this internal circuit and no current will flow in the armatures of the machines beyond the ordinary load currents*. The vector diagram shown in Fig. 113 refers to the internal circuit of the machines and E_1 and E_2 represent the pressure vectors when the machines have the correct phase relationship. Now suppose that owing to some cause or other, say a cyclical speed variation of one of the prime movers, the pressure of machine number two falls a little behind true opposition in the internal circuit and its pressure vector takes up the position E_2 as shown in the figure. There will now be a resultant pressure, represented by R, in the local circuit of the two machines, and this, though perhaps small

---

* It is assumed that the pressures given by the two machines are numerically equal.

in amount, may send a considerable current through the two armatures since the impedance in the circuit will be quite low and, further, this current will lag considerably behind the resultant pressure since the impedance of the local circuit will chiefly consist of reactance rather than of resistance. The local current is shown by the vector OI and it is necessary to point out that this is not the only current flowing in the armatures of the two machines; it should be regarded as superposed on whatever current each armature is supplying to the load circuit. This local current is, at any rate, within 90° of the pressure E_1 generated by alternator number one, and, as far as this alternator is concerned,

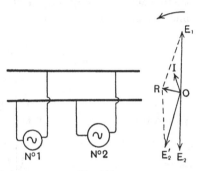

Fig. 113.

is a generating current; this alternator is then, owing to the local current, giving out electrical power over and above what it is supplying to the external circuit. Further, the local current is within 90° of being in phase opposition to the pressure E_2 given by alternator number two, and, as far as this machine is concerned, may be regarded as a motoring current, that is this machine is, on account of the local current, absorbing electrical power. Summing up we see that the effect of the local current is to add load to the alternator which tends to gain relatively to the other, and to drive the alternator which tends to run too slowly, or in other words the effect of the local current is to tend to pull the machines into their correct phase relationship again by loading up the leading machine and taking load off the machine which is running too slowly. In order to obtain good parallel running qualities in machines it is desirable that a small phase displacement shall produce a large local current and this will be attained by having the armatures with small impedance; further, it is of equal importance that the ratio of reactance to resistance shall be high in order that the local current shall lag as much as possible behind the resultant pressure. If the local current is nearly in phase with the resultant pressure it will be nearly wattless when considered in conjunction with E_1 and E_2 and so will not exert any great synchronising power.

It will be of interest at this stage to examine the problem of alternators connected in series, in order to see if there is any corresponding interaction between the machines in this case tending to pull them back into step should their correct phase relationship become disturbed from any cause. Here there will

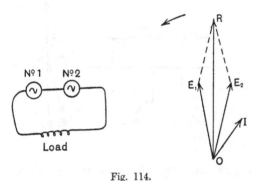

Fig. 114.

be no purely local current and the only current flowing through the armatures of the machines will be the load current; the vector diagram shown in Fig. 114 refers to the external circuit. When the machines are in correct phase relationship the two pressures will be exactly in phase with each other; now suppose that machine number two lags a little, then the phase relationship will be as shown in the figure in which E_1 and E_2 represent the pressures of the two machines and R the resultant pressure tending to send current round the external circuit.

Let us suppose that the load is slightly inductive, the usual case, the load current will then be represented by the vector OI. We now have two machines giving equal pressures and sending the same current but the angle of phase difference between pressure and current will be rather less in the case of the machine that is slowing down than in the case of the other machine, that is to say, the load on the machine which is tending to slow down is increased while the load on the other machine is decreased, thus the interaction between the machines, so far from pulling them into step again, will tend to pull them more and more out of step. If by any chance the load took a leading current a little thought will show that there will then be a slight tendency to pull the machines into step, since the machine tending to lead would then have more load placed upon it and *vice versâ*.

The above investigation demonstrates that if it is desired to run alternators in series they must be mechanically coupled in a very rigid manner. Reverting now to the case of machines intended for parallel running, it will be realised that the following conditions must be satisfied before they are paralleled:

(*a*) The frequencies of the pressures given by the two machines must be equal.

(*b*) The magnitudes of the pressures given by the two machines must be equal or practically so.

(*c*) The two pressures must be in phase opposition as regards the local circuit composed of the two machines.

It is also desirable that the wave forms of the two machines be similar, otherwise, even if correct phase relationship is obtained, the two pressures cannot balance each other round the local circuit at every instant and thus a local current is bound to flow giving rise to undue heating.

The equality of the magnitude of the pressures can be ascertained by voltmeters connected in the usual way, while the equality of frequencies and the moment of correct phase relationship can be ascertained by what is known as a synchronising device or, in its more elaborate forms, a synchroscope. The simplest type of synchronising device consists of a pair of lamps, connected across the switches of the incoming machine in such a way that when the switch is closed the lamps are short circuited by the two blades of the switch as shown in Fig. 115. Before the switch is closed the pressure across the lamps will be the resultant of the pressure across the bus bars and the pressure of the incoming machine and, if the frequency of the latter is rather higher or lower than the frequency of the pressure on the bars, the pressure on the two lamps (which are in series) will vary periodically from zero (when the bar and machine pressure are in opposition in the local circuit) to double the bus bar pressure (when the bar and machine pressure are in phase in the local circuit). Thus the brilliancy of the lamps will vary with a frequency which will depend upon the difference in the frequencies of the two pressures and by adjusting the speed of the incoming machine the difference between the two frequencies may be made quite small. The correct moment to close the switches (having first seen that the two pressures are practically equal) is when the pressures are in opposition as regards the local circuit, that is when there is no pressure across the switch and the lamps are dark. The arrangement described above does not give a very sensitive indication as to the precise moment for

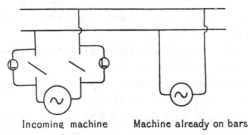

Incoming machine   Machine already on bars

Fig. 115.

synchronising, since the lamps will be dark for quite an appreciable time on each side of the moment of zero voltage, thus the switch may be closed when there is quite a considerable voltage across it and yet not enough to render the lamps visible; further, a burnt out lamp might result in the switch being closed at the wrong moment. A much better arrangement is to connect the lamps as shown in Fig. 116, in which case, as far as the circuit through the lamps is concerned, one alternator is reversed relatively to the other and thus the correct moment for closing the switch is at the moment of maximum brilliancy of the lamps, a moment which is readily observed owing to the rapid variation of the candle power of a carbon filament lamp with rise of pressure.

A voltmeter, of a type which is suitable for use on alternating current circuits and which is fairly dead beat, could be substituted for the lamps as a synchronising device, and the connections could be arranged to synchronise either with the reading a maximum or zero as desired.

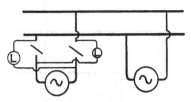

Fig. 116.

For high pressure circuits it will be necessary to connect the synchronising lamps across the secondaries of special small transformers as indicated in Fig. 117 (*a*) and (*b*). When the two magneto-motive-forces help each other in driving flux through the limb of the transformer round which the secondary is wound the lamps will light and *vice versâ*, and it is obviously a simple matter to arrange to synchronise either with lamps light or with lamps dark as may be desired.

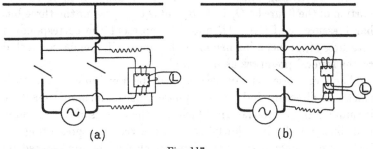

(a)　　　　　　　　　　　　　　(b)

Fig. 117.

Three phase low pressure machines can be synchronised with lamps dark by using the connections shown in Fig. 118.

All the devices described up to the present tell whether the difference in frequency between the incoming machine and the machines already on the bars is great or small, but they afford no indication as to whether the incoming machine is fast or slow, and this knowledge is desirable since it often means that time can be saved in synchronising; to remedy this defect devices which are known as rotary synchronisers have been introduced.

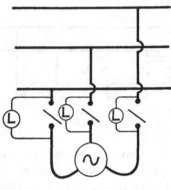

Fig. 118.

The rotary lamp synchroniser. Consider a three phase alternator with three synchronising lamps connected as in Fig. 119 (*a*), that is to say, with one lamp connected directly across the switch belonging to one phase and with the other two lamps crossed in regard to the two remaining phases. In the vector diagram in the lower portion of the figure let O_1A_1, O_1B_1 and O_1C_1 represent the three bus bar pressures and OA, OB and OC represent the corresponding pressures of the incoming machine, the lamps will then be connected to the several pressures as indicated.

Each of these sets of vectors will be revolving in an anti-clockwise direction and let us suppose that the incoming machine is running slightly too fast, in which case the vectors of this machine will continuously, but slowly, gain on the vectors representing the bus bar pressures, a state of affairs which may be represented by thinking of the bus bar vectors as stationary and the machine vectors rotating slowly in an anti-clockwise direction. At the instant shown in the diagram we see that the pressures OA and O_1A_1 exactly oppose each other and consequently the lamp L_1 will be dark, but the pressures OB and O_1C_1 will have a resultant value (having of course a frequency equal to the frequency of the machine pressure) and consequently the lamp L_3 will be bright (though not at its brighest), a state of affairs which will also apply

to the lamp L_2. As the machine vectors move in an anti-clockwise direction relatively to the bus bar vectors they will soon gain
120°, and then OC will become parallel to O_1B_1 with the result
that the lamp L_2 will become dark the others being illuminated:
a further gain of 120° will result in OB becoming parallel with O_1C_1
and the lamp L_3 will then be dark the others again being bright
(though not at their maximum brightness).   If we have the three

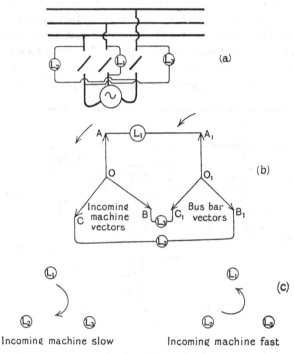

Fig. 119.

lamps arranged in the form of a triangle the position of maximum
brightness and the position of darkness will appear to rotate in an
anti-clockwise direction. If now the incoming machine be regarded
as running too slow, a mode of investigation similar to that used
above will show that the position of zero illumination will revolve
round the triangle in a clockwise direction and we therefore have
a simple method of ascertaining whether the incoming machine
is running too fast or too slow. The correct moment to close the

switch is when the lamp L_1 is dark. The simple arrangement described above is only suitable for low pressure machines but devices working on the same principle, with the lamps operated through transformers, have been used on high tension alternators of large size.

The Weston synchroscope. A very ingenious and simple form of synchroscope which, in addition to indicating when the machines are in phase, also indicates whether the incoming machine is fast or slow, is made by the Weston Electrical Instrument Co., Ltd., to whom the author is indebted for information concerning the working of the instrument and also for Fig. 120 which shows the internal connections. In order to examine the operation of the instrument we may regard it as consisting of two parts:

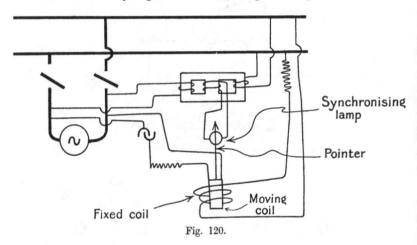

Fig. 120.

(a) The portion comprising the lamp transformer and the lamp. This in itself is a type of synchroscope since the lamp is connected so that when the machine and bus bar pressures oppose each other in the local circuit the lamp is bright, and when the two pressures help each other the lamp is dark, thus the lamp will flicker more and more slowly as the two frequencies become equal, the correct moment for synchronising being when the lamp is bright.

(b) A portion which is practically a wattmeter movement comprising fixed and moving coils, the moving coil being connected across the terminals of the incoming machine in series with some resistance and a condenser, and the fixed coil, which possesses considerable inductance, being connected across the bus bars in series with a resistance. The moving coil is spring controlled so that its zero position is upright in the figure.

The relative values of the resistance, inductance and capacity are so chosen that when the bus bar and machine pressures are in opposition (in regard to the local circuit through the machines) the two currents sent through the instrument will be 90° apart with the result that the pointer will be at zero. There will also be no deflection of the pointer when the machine and bus bar pressures are in phase in regard to the local circuit, but for every other relative phase of the machine and bus bar pressures there will be a deflection one way or the other and this deflection will be a maximum when the two pressures are 90° apart. Thus, if the pointer is observed, it will be seen to oscillate more and more slowly as the frequencies of the two pressures become more nearly equal, always making one complete oscillation for each cycle gained by one pressure on the other. The reader should be clear that it will not be safe to synchronise when the pointer is at zero since this will occur not only when the machines are in opposition in the local circuit but also when they are in phase.

The method of indication as to whether the incoming machine is too fast or too slow can be realised if the two operations which are described above are considered in conjunction with each other. As a matter of fact, the pointer of the wattmeter movement is arranged to move behind a ground glass screen and what is really seen is the shadow of the pointer which is thrown by the synchronising lamp, and this of course will only be seen during the period when the lamp is bright. Now the lamp becomes bright only once during each period gained or lost by the incoming machine and during this time the pointer will have made one complete oscillation, further, the lamp will have its maximum brightness at the time during which the pointer is passing through its zero position. A little thought will show that the shadow of the pointer will always be seen moving in one direction because when the pointer is moving in the reverse direction the lamp will be dark.

The appearance observed may be likened to the spokes of a wheel passing across the ground glass thus giving a distinct rotating effect which, in the actual instrument, is clockwise when the incoming machine is fast. A similar appearance will be observed if the incoming machine is too slow, but this time it will be the reverse swing of the shadow of the pointer which will be observed, the effect being that of the spokes of the wheel moving in the opposite direction.

The points mentioned above will be clearly seen from Fig. 121, in which the vector representing the bus bar pressure is supposed to be vertical and the different relative positions of the machine

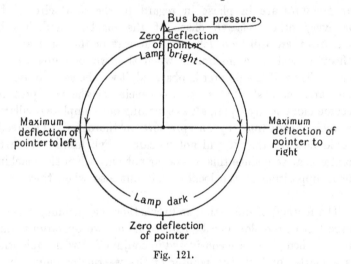

Fig. 121.

vector are shown round the circle, the state of the lamp and pointer being shown for each position of the latter.

The correct moment for synchronising is when the shadow of the pointer is clearly visible, is in the upright position, and is stationary or moving slowly. The indications of the device are not quite independent of frequency but are very nearly so over the variations which are likely to occur in central station work; it is, however, customary to calibrate the instrument for the normal frequency and pressure on which it is to be worked. The instrument is directly applicable for single phase machines, and can be used across two lines of three phase machines, due care being taken when connecting up that the bus bar and machine leads are correctly paired.

Motor type rotary synchroscope. This type of synchroscope, which is manufactured by Messrs Everett, Edgcumbe and Co., Ltd., to whom the author is indebted for information and for Figs. 122, 123 and 124, is not only of interest on account of its large commercial use, but also on account of the scientific principles involved.

The device essentially consists of a small motor provided with fixed and moving windings, the connections of which are shown in Fig. 122.

The stator of this motor is wound with a single phase winding and is connected to the terminals M_1 and M_2 which are also connected to the leads of the incoming machine, transformers being used instead of a direct connection when the pressures are high. The rotor is wound for two phase currents and is connected across the bus bars, the two phases being obtained by the phase splitting device indicated in the figure, from which it will be seen that a choker is inserted in series with one winding, and a resistance (which is really a lamp used to illuminate the dial) in series with the other winding. These two coils are rigidly joined together, but, since the currents are conveyed to them by slip rings, are quite free to rotate as a whole. It is clear that the fixed coil will produce an alternating field, while the two moving coils will produce a rotating field.

The arrangement is in fact exactly similar to that used in the power factor indicator described on page 123 with the minor exceptions that the coils are fed from two machines instead of from a single circuit, and that the moving coils are free to rotate continuously.

As in the case of the power factor meter the moving system will take up a position such that the axes of the alternating and rotating fields will coincide at the instant the former is passing through its maximum value. If the incoming machine is in opposition to the bus bar pressure (as regards the local circuit) the pointer takes up a vertical position and will be stationary if the frequencies are equal; if the incoming machine is a little out of correct phase the pointer will take up an inclined position, but will still be stationary if the frequencies are equal. If the frequencies of the incoming machine and the bus bar pressures

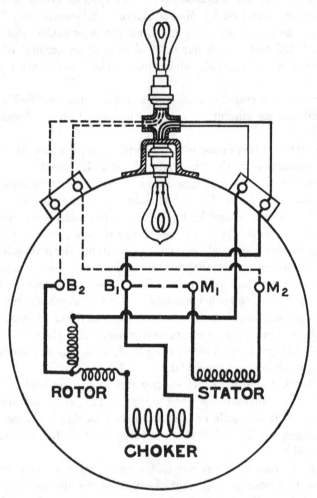

Fig. 122. Synchronising Lamp.

are not equal, the phase relations of the two machines will be continually changing and this will result in a slow rotation of the pointer one way or the other, depending on whether the incoming

machine is too fast or too slow, the correct moment for synchronising being when the pointer is vertical and at rest (or moving quite
slowly).

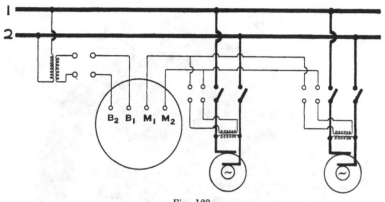

Fig. 123.

An interesting auxiliary device is fitted to the instrument
under consideration in order that the indications may be visible
at the position occupied by the alternator, which will often be
some considerable distance away. The pointer is arranged so that
by means of a clutch it can move a light frame, carrying red and
green glasses, in either direction, the motion of the frame being
limited by stops to an angle of about 30°. The glasses move in
front of a small opening in the dial of the instrument behind which
is placed the lamp previously referred to, thus, when the incoming
machine is going too fast and the pointer is revolving in one
direction, the red glass is brought in front of the lamp, and when
the incoming machine is too slow the green glass is brought in
front of the lamp. Again, if the switchboard is not visible from
the prime mover, the motion of the frame may be used to close
auxiliary circuits which light red or green lamps, depending on
the direction of motion of the pointer, which are situated adjacent
to the prime mover. Fig. 123 shows the external connections of
the synchroscope to the machine and bus bars, transformers being
made use of, since the connections are intended for high pressure
circuits, and a plug device is used to allow of the synchroscope
being used for any one of several alternators.

Fig. 124 shows a view of the instrument with the cover removed, the various parts being lettered as in the schedule below:

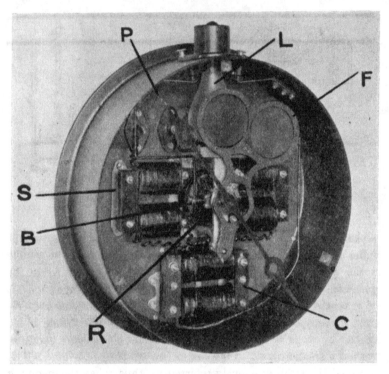

Fig. 124.

S Single phase stator fed from incoming machine.

R Two phase rotor fed from bus bars.

L Lamp placed in series with one of the rotor windings and which also serves to illuminate the coloured glasses.

C Choker placed in series with the other rotor winding.

P Pointer attached to rotor.

B Brush gear to convey currents to rotor.

F Frame carrying the red and blue glasses.

An ordinary synchronising lamp can be fitted, if required, as shown in Fig. 122, and for two and three phase circuits precisely the same instrument is used but it is of course connected across two lines only and, in installing the synchroscope for use under these circumstances, care must be taken that the correct lines of the machine and bus bar are connected to the two sides of any one pole of the main switch.

EXAMPLES

1. Make a developed diagram of the armature winding of a single phase alternator having twelve slots per pole, of which eight are made use of, and using as many coils as there are poles.

2. Repeat the above example for the case where half as many coils are used as there are poles.

3. Arrange a three phase winding for the core described in example (1) making use of two shapes of end connections and utilising four slots per pole per phase.

4. Repeat the last example using three shapes of end connections and arranging that it is possible to split the armature without disturbing any of the coils.

5. An alternator having sixteen poles is run at a speed of 375 revolutions per minute. If there are 200 conductors in series per phase and the useful flux per pole is 4.5×10^6 lines, calculate the R.M.S. value of the pressure produced per phase. Assume the Kapp coefficient to be 2·22.

Answer. 1000 volts.

6. What must be the flux per pole in the above machine in order that a pressure of 3750 volts per phase may be developed?

Answer. 16.87×10^6 lines.

7. The generated pressure of a single phase alternator is 180 volts, the current being 30 amperes. If the resistance of the armature is ·4 ohm and the reactance 1·2 apparent ohms, calculate the pressure at the terminals of the machine if the current is in phase with the generated pressure.

Answer. 172 volts.

8. Repeat the above example if the current lags 60° behind the generated pressure. *Answer.* 143 volts.

9. Draw a developed diagram of a four pole, three phase, armature winding, showing the correct inter-connection of the three phases if they are starred. For simplicity take one armature conductor per pole per phase.

10. Repeat the above example for the case when the phases are meshed.

CHAPTER VIII

STATIC TRANSFORMERS

We have already seen in Chapter III that when an alternating magnetic flux is produced in an iron cored choker, by the passage of an alternating current in a coil encircling the core, an induced pressure is produced in the coil owing to the effect of self-induction, the induced pressure being in phase opposition to the applied pressure if the resistance of the coil is negligible. Induced pressures may also be produced in adjacent circuits if these are in such a position that a change of current in the first, or primary coil, causes a change in the total flux linking with the second, or secondary coil. Such pressures, produced in circuits adjacent to that in which the primary current is flowing, are said to be due to mutual induction though their mode of production is exactly similar to pressures produced by self-induction. In Fig. 125 let P be a primary circuit through which an alternating current is flowing, then, if the secondary circuit is in the position shown at (a), any change of current in the primary will result in a change of flux linking with the secondary, that is to say the two circuits will have considerable mutual induction; on the contrary, if the secondary circuit is arranged as shown at (b), change of current in P will not produce a change in the total flux linking with S and the two circuits will then have no mutual induction. The units used in connection with mutual induction are similar to those already used in connection with self-induction and the definitions given for the unit of self-induction may be applied to the unit of mutual induction if due care is taken to indicate that the current, or change of current, producing the flux, or change of flux, is in a neighbouring circuit to that in which the linkage is considered.

Thus, for example, two circuits are said to have a mutual inductance of one henry if their arrangement is such that a rate of rise of current of one ampere per second in the primary causes an induced pressure of one volt in the secondary. The phenomena occurring in two circuits having mutual induction do not, as a rule, simply consist of the production of an induced pressure in the secondary. If the secondary is open circuited so that no current can flow within it, it is true that no further action takes place, but if, as is usually the case, a secondary current flows, it will react on the primary circuit.

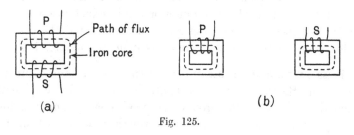

Fig. 125.

The phenomena occurring under these conditions are perhaps best investigated by examining the matter from an experimental point of view, and for this purpose let us consider the circuits shown in Fig. 126. When the secondary circuit is open the

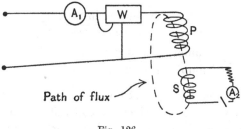

Fig. 126.

readings on the ammeter and wattmeter in the primary circuit will be small, but when a secondary current is allowed to flow the readings of the primary instruments will increase, and, further, any increase in the secondary current will result in an appropriate increase in the readings of the primary instruments.

If the coils are arranged so that the mechanical force between them can be measured, it will be found that there is repulsion between them when secondary current flows, and that this force of repulsion increases with the magnitude of the secondary current.

The static transformer, so called because it has no material moving parts, is a device in which two coils are arranged to have very considerable mutual induction, and when the primary is supplied with alternating pressure at a given frequency we have produced in the secondary another alternating pressure at the same frequency, but which may have either a higher or a lower magnitude. In order to obtain the maximum mutual inductance with the minimum expenditure of copper it is usual to wind the coils on a closed and laminated iron circuit as indicated in Fig. 125 (*a*).

Pressure ratio in transformers. Using the arrangement indicated in the last-mentioned figure, there will be very little magnetic leakage if the iron is of good quality and the flux density be kept low. Every turn, whether it is part of the primary or part of the secondary, will be cut by practically the same flux and will have the same value of induced pressure produced within it. The relative values of the total induced pressures in the two windings will therefore simply depend upon the relative number of turns in the two windings, and we shall have

$$\frac{\text{Pressure induced in primary}}{\text{Pressure induced in secondary}}$$

$$= \frac{\text{Number of primary turns}}{\text{Number of secondary turns}} = \frac{T_{pb}}{T_s}.$$

Further, if the resistance of the primary winding be neglected, the applied pressure in the primary circuit will be numerically equal to the induced pressure and so we have very approximately

$$\frac{\text{Pressure applied to the primary}}{\text{Pressure induced in the secondary}} = \frac{T_p}{T_s}.$$

In practice the induced pressure in the secondary circuit will be rather less than that indicated by the above formula owing to the effect of resistance of the windings and to the effect of magnetic leakage, and the departure from the value indicated by the formula will be greater the greater the load.

Consideration of ideal transformer. Consider the case of a transformer whose windings have no resistance, the magnetic leakage and the core loss also being negligible. If an alternating pressure is applied to the primary, a magnetising current will be taken by the transformer of such an amount and phase that the resulting core flux will produce a back pressure in the primary winding which will be numerically equal to the applied pressure but will be opposite to it as regards phase. The primary winding will behave like a choking coil, that is to say, the magnetising current and core flux will be in phase with each other and will both be 90° behind the applied pressure. The induced pressures, in both the primary and secondary, will be 90° behind the core flux and will of course be in phase with each other though they may have very different magnitudes, depending upon the relative number of turns in the primary and secondary windings; the complete vector diagram for the ideal transformer when on no load is shown in Fig. 127.

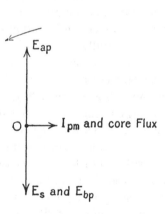

Fig. 127.

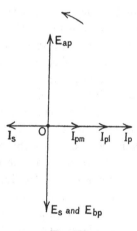

Fig. 128.

OE_{ap}	Primary applied pressure
OI_{pm}	,, magnetising current
OE_{bp}	,, back pressure
OE_s	Secondary pressure

OI_s	Secondary current
OI_{pl}	Primary load current
OI_p	Total primary current
Other symbols as in Fig. 127	

Next let us suppose, while still thinking of the ideal transformer, that the secondary circuit is closed on a purely inductive load. A secondary current will flow, which will be 90° behind the secondary pressure, and this current will of course tend to exert a magnetic effect on the core. We see from the vector diagram shown in Fig. 128 that the magnetic effect of the secondary current will be directly opposed to the magnetic effect of the primary magnetising current, and there will thus be a tendency to diminish the core flux; but, if any weakening of the core flux actually occurred, the induced pressure in the primary circuit would no longer completely balance the applied pressure, and so an additional primary load current would flow in the primary circuit whose magnitude and phase was such as to neutralise the magnetic effect of the secondary current, thus leaving the primary magnetising current free to produce the required value and phase of the core flux. The reader should realise that at all loads the core flux, which is due to the resultant effect of the total primary and secondary currents, must have a certain definite value which is the value taken up at no load. The ampere turns of the primary load current will be numerically equal, but of opposite phase, to the ampere turns of the secondary current, and the total primary current will be found by adding the load and magnetising components together; in the particular case under consideration they are in phase and so will be added arithmetically.

Let us now suppose that the secondary circuit is composed of a non-inductive resistance, the secondary current will then be in phase with the secondary pressure but will again tend to disturb the core flux. The primary magnetising current and the secondary current will be at right angles and, considering their vectoral resultant, we see that the tendency of the secondary current will be to increase the core flux, but it will also disturb the phase of the flux and again, as in the previous case, the flux will no longer be competent to produce the required back pressure in the primary circuit. The presence of the secondary current necessitates therefore the flow of an additional load component of the primary current, and this will be of such a magnitude and phase as to neutralise the magnetic effect of the secondary current. The primary magnetising current and the primary load current will now differ in phase by 90° and the total primary current will be their vectoral resultant (see Fig. 129).

The most general type of load on the secondary circuit will be one in which the current lags behind the pressure by some angle between 0° and 90° and will be intermediate in character between the loads in the two cases already considered; the phenomena occurring will be similar in character to those already discussed, the appropriate vector diagram being shown in Fig. 130.

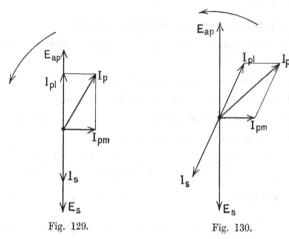

Fig. 129.   Fig. 130.

Relationship between primary and secondary current in a transformer. We have already seen that the ampere turns of the primary load current are numerically equal to the ampere turns of the secondary current, and we have therefore

Primary load current × Primary turns

$$= \text{Secondary current} \times \text{Secondary turns},$$

or $\qquad \dfrac{\text{Primary load current}}{\text{Secondary current}} = \dfrac{T_s}{T_p}.$

At, or near, full load, the primary total current is not very different from the primary load current, since the magnetising current is usually comparatively small in amount and, further, is usually at an angle approaching 90° from the load current, and so very approximately we have

$$\frac{\text{Primary current}}{\text{Secondary current}} = \frac{T_s}{T_p}.$$

J. C. 13

Again, a study of the vector diagrams will show that the phase difference between current and pressure is, for non-inductive or partially inductive loads, rather greater for the primary than for the secondary circuit owing to the effect of the magnetising current; the difference is not great for ordinary transformers operating at, or near, full load and generally we may say that a small phase difference between current and pressure, and therefore a good power factor, in the secondary will lead to a like state of affairs in the primary.

Pressure generated in a transformer winding.

Let N represent the number of turns in one winding of a transformer,

A the cross-sectional area of the core,

B the maximum flux density in the core,

f the frequency of the applied pressure in cycles per second,

and E the pressure produced in the coil.

Now the lines cut per turn per cycle will be $4BA$, and the lines cut per turn per second will be $4BAf$, therefore the average pressure per turn will be $\dfrac{4BAf}{10^8}$, and the average pressure per coil will be $\dfrac{4BAfN}{10^8}$ volts. If we assume the pressure wave to be sinusoidal, the ratio between the R.M.S. and average values of the pressure will be 1·11, and thus we see that the R.M.S. value of the pressure produced per coil will be

$$\frac{4 \times 1\cdot11\ BAfN}{10^8} = \frac{4\cdot44\ BAfN}{10^8} \text{ volts.}$$

If the pressure does not follow a sinusoidal wave the constant will be slightly different, being greater for peaked waves and less for flat-topped waves.

Construction of Modern Transformers.

The general arrangements of transformers as manufactured by different makers vary considerably, but two main types may be distinguished:

(1)   The core type, and

(2)   The shell type.

In the former type a rectangular core is provided round which the primary and secondary coils are wound and generally we may say that in the core type the copper surrounds the iron. Typical three phase transformers of the core type are shown in Figs. 137 and 138.

In the shell type, on the contrary, the iron may be said to surround the copper coils, stampings being used with holes punched in them through which the windings are passed. Typical examples of the shell type are shown in Figs. 139 and 141. Both types are largely used in practice and there is probably little to choose between them, the core type would seem to be somewhat preferable from the point of view of ease of repair, while the shell type offers greater mechanical protection to the windings.

The core.   In arranging a transformer core several points must be considered, these being as follows:

(1)   The losses taking place in the core must be kept low otherwise the efficiency will be low, especially at low loads. Since the core losses depend largely on the flux density used (eddy current losses increase as B^2, and hysteresis losses as $B^{1\cdot6}$) low flux densities are desirable and we find values from 4000 to 9000 lines per square cm. commonly used. It is hardly necessary to mention that lamination is resorted to for the purpose of keeping down the eddy current loss, common thicknesses of the laminations being from 10 to 20 mils. Again, it is desirable to use an iron for a transformer core whose loss, when subjected to an alternating flux, is inherently low, and from this point of view the so-called alloyed irons are particularly useful. Perhaps the most widely used iron is that known as Stalloy, and this possesses the double advantage of having a low hysteretic constant, and also a low inherent eddy current loss owing to its high specific resistance. The comparative losses occurring in ordinary iron and in Stalloy are shown in Fig. 131, from which the advantage to be obtained by using the latter is clearly seen. For information as to the losses in Stalloy the author is indebted to the manufacturers, Messrs J. Sankey and Co., Ltd., of Bilston.

Stalloy is also remarkably free from deterioration due to "ageing," even when used at so high a temperature as 100° C.

(2) The magnetising current required by the transformer should be as small as possible since a large magnetising current will diminish the power factor at all loads and more particularly at small loads. To obtain a low magnetising current it is necessary that the reluctance of the path traversed by the useful flux of the transformer shall be as low as possible and this is to some extent secured by working the iron at low flux densities thus securing a high permeability. Modern transformers are invariably constructed with former wound coils and in order that these may be used it is necessary to have at least one joint in the magnetic circuit and this, unless very carefully designed and constructed,

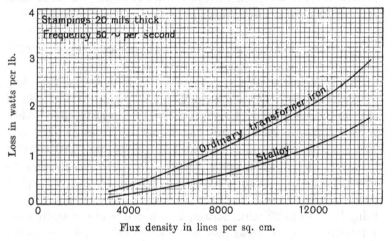

Fig. 131.

may add very considerably to the reluctance of the path traversed by the flux. In any one layer of the core stamping the joint is simply one of the butt variety, and the magnetic lines when passing this point have to pass through a small air gap. It is not essential however that the joint, or joints, in the core, considered as a whole, be made by simply butting two sets of stampings together, it is possible to arrange the stampings so that in successive layers the joints in the stampings break step thus giving, in the core as a whole, what may be described as an interleaved joint. Such an interleaved joint may be made by having the joints in one layer

of stampings as shown at (a) in Fig. 132, the joints in the next layer
being as shown at (b), and so on in alternate layers.   Such a joint
will not be so convenient from the constructional and repair points
of view but will certainly give a better magnetic circuit than if the
joint were a simple butt one.   It should be noted that the interleaved joint does not do away with the necessity of the magnetic
lines having to pass through an air gap, but it does away with the
necessity of the lines having to pass through the non-magnetic
gap at a high flux density.

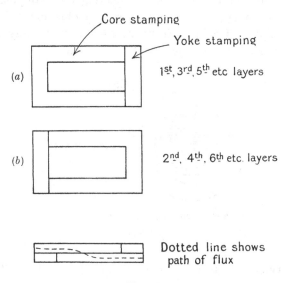

Fig. 132.

(3)   The length of each turn of the winding should be kept as
small as possible so far as is compatible with condition (1) stated
above.

To help in this matter the flux density in the core is sometimes
made rather higher than in the yoke portion of the magnetic
circuit, and stepped cores may also be used in large transformers.
A stepped core is one in which two or more breadths of stampings
are made use of, the idea being to obtain as great an area
of iron as possible in a circular turn of given diameter (see
Fig. 133).

(4) It is essential that the stampings forming any one core
and yoke be firmly secured together and, further, if the magnetic
circuit is made of two parts, for convenience in erection, the two
parts should be firmly clamped or otherwise secured together.
If this consideration is not attended to vibration and humming
will result (due to the appearance and disappearance of magnetism
owing to the alternating currents) and the vibration may ultimately
cause abrasion of the insulation of the conductors. At the same
time it is essential that the means of securing the various parts
together shall not result in short circuiting the stampings or in
any way forming short circuited paths embracing alternating
magnetic fluxes, for in such event, what is practically a short
circuited secondary winding would be formed.

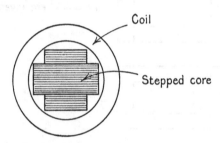

Cross-section of stepped core

Fig. 133.

In very small transformers the stampings may be well bound
together with tape, wooden dowel pins being used, if considered
necessary, to keep the stampings from moving relatively to each
other; there would be a grave risk with this construction, however,
of the tape being cut through by vibration thus loosening the
plates. A better method of clamping the plates is to use stout
end plates, preferably of bronze or brass, and to clamp the core
plates tightly between these by means of steel bolts passing through
insulated tubes and also insulated from the end plates by insulating
washers. As a further precaution against the possibility of the
production of eddy currents the bolts should be put in such
positions that the lines joining them run parallel to the magnetic
lines (see Fig. 134).

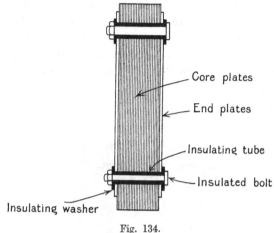

Core plates

End plates

Insulating tube

Insulated bolt

Insulating washer

Fig. 134.

If the plates to be bolted together are short (as in the case of the yoke portions) the stout end plates may be made longer than the core plates proper, the clamping then being done by external bolts as in Fig. 135.

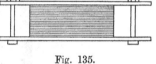

Fig. 135.

The yoke portions of the magnetic circuit may be clamped on to the core portions as indicated in Fig. 136, which has been kindly furnished to the writer by the Brush Electrical Engineering Co., Ltd., of Loughborough, along with the material used in the preparation of Figs. 137, 138 and 140.

For three phase working, three single phase transformers can be used if desired, but special transformers having three cores are also largely made use of, the winding of one phase being placed on each core. Most commonly the three cores are joined by a single yoke, as in Fig. 138, but a triangular arrangement of the cores may be used if desired (see Fig. 137). The latter method is likely to give rise to a more expensive construction but a better magnetic balance in the cores is attained.

Windings of transformers. In arranging transformer windings it is obviously very necessary to secure adequate insulation between the primary and secondary coils, and this will be of particular importance if the primary carries high tension and the secondary low tension for distribution to lighting consumers. Good insulation between both windings and earth will also be necessary. Again, in all transformers a certain amount of magnetic leakage takes place due to the fact that the windings, in addition to setting up a useful flux which links with both, also set up leakage fluxes which link with one winding only. These leakage fluxes give rise to

Fig. 136.

leakage reactance of the windings, which is an important cause of pressure drop, and hence it is desirable to keep the leakage fluxes low. It is perhaps not desirable to have the leakage reactance too low since its presence limits the magnitude of the current which flows in the event of a short circuit on the secondary side. As a rule it is more important to keep down the leakage in lighting transformers than in transformers for power loads.

These conditions are conflicting since separation of the windings will be useful in order to obtain good insulation, while it will be desirable to bring the windings in close proximity in order to keep down magnetic leakage. Types of transformer coils are shown in Figs. 138 and 141 ; in the former all the coils are of equal

Fig. 137.

mean diameter and are short in length, both primary and secondary windings being split up into a number of sections, and on any limb primary and secondary sections are placed alternately, the winding being completed by connecting all the primary sections in series and all the secondary sections also in series. This arrangement has sometimes been called the disc type of winding. In the

second figure mentioned above the different sections into which
the primary are divided are much longer and are of different
diameters and they are arranged to fit within one another or co-
axially.

The different sections of the coils are usually separated from
each other and from the core by distance pieces so that free

Fig. 138.

circulation of air or oil is allowed for cooling purposes. Occasionally
in lighting transformers an earth connected metallic shield is
inserted between the primary and secondary windings as a protection against the risk of the high pressure penetrating into the
low pressure winding; care must be taken that the shield does
not form a short circuited secondary and this is prevented by
slotting. Some account of the insulation adopted in a transformer

is given on page 206 and it may be remarked that the end turns of high tension transformers need particularly careful insulation since they may be subject to severe stresses during switching operations, or when a sudden change of terminal pressure takes place due to any other cause, owing to what has been termed concentration of potential. This phenomenon arises from the fact that each turn of a transformer winding possesses both inductance and capacity (to earth for example) and thus, when the pressure of the end turn is suddenly raised by closing a switch, a charging current is taken by each turn. Before the potential of the second turn can rise, the necessary quantity of electricity for charging it must pass through the first turn and this takes an appreciable time owing to the inductance of the first turn. The above remarks apply to each of the turns near the end of the winding and we see that before the pressure can distribute itself through the winding a certain time must elapse, and during this time there will be an unduly high pressure between adjacent turns of the first part of the winding which may cause a breakdown of the inter-turn insulation*.

The trouble may be minimised by placing a well insulated choking coil between the switch and the transformer (the inductance being of such a value that there is no appreciable pressure drop under normal conditions) or by additional insulation on the end turns of the transformer windings.

Careful insulation of the leads of high tension windings is also necessary, this being usually effected by porcelain bushings.

Cooling of transformers. One of the most important considerations in connection with transformers is that of heating, and the difficulties of dealing with this question increase rapidly as the size of the transformer increases. If we assume that there will be approximately the same amount of energy to be dissipated per unit volume of working material of the transformer in all sizes, we see that the total amount of heat to be dissipated will vary as the volume of the transformer, that is, as the cube of the linear dimensions; while for similar constructions the area available for the dissipation of the heat increases as the square of the linear dimensions, and this argument bears out the above statement.

* See paper by Mr Peck, *J. I. E. E.*, Vol. 40, p. 499.

For small sizes and moderate pressures (say up to 3000 volts) natural air cooling may be used. In this case the transformer may be placed in a metal containing case to which air has free access, or it may be placed in a completely closed case made of cast iron or sheet steel or a combination of the two. If the latter method is used it should be noted that the heat to be dissipated is first given to the air in the case, then to the case itself, and finally to the outside air. For large sizes of air immersed transformers it may be necessary to make use of a forced draught in order to obtain efficient cooling and the air used may either be blown directly on to the coils and iron and through the ventilating ducts, or it may be blown on to the external case; the former method will be the more efficient but in many cases it would be necessary to filter the air before passing it into the transformer.

There are great advantages to be obtained by immersing the transformer in oil; the oil not only minimises the risk of failure of insulation but it is also a much better medium for the conduction of the heat away from the working parts of the transformer than air. Needless to say the oil used should have a high flash point, high insulating properties, be as free from moisture as possible and also from acid and, further, should not cause any deposit in course of long use at a considerable temperature. It is also very necessary that the case be oil tight, a condition not easy to attain. Such oil immersed transformers using natural cooling can be made up to the highest pressures and in sizes up to say 1000 K.W., though in the larger sizes corrugated cases must be used in order to obtain the necessary area of cooling surface to the external air. In the largest sizes of transformers it is usual to cool the oil by artificial means using either a spiral tube in the upper part of the case, through which a stream of cold water is continually passed, or extracting the hot oil from the upper part of the case, cooling it externally to the transformer, and returning it to the lower part of the case. It has sometimes been thought that the presence of oil would add to the fire risk of the installation, but this fear has hardly been borne out in practice.

The photograph shown in Fig. 139 represents a shell type transformer as made by the British Westinghouse Manufacturing Co., Ltd., to whom the writer is indebted for permission to make

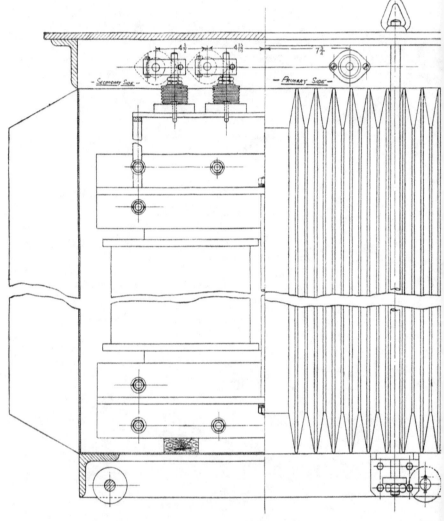

- SECONDARY SIDE -   - PRIMARY SIDE -

Fig.

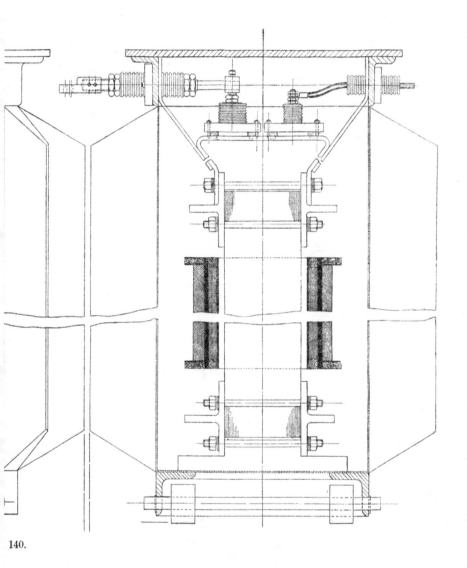

140.

use of the illustration: it is of the oil immersed type and the plain containing case is seen alongside.

Complete general arrangement drawings of a single phase core type transformer are shown in Fig. 140, which illustrates a 137 K.V.A. transformer working from 6000 to 400 volts at a frequency of 50 cycles per second.

Features of particular interest, since they are not shown elsewhere, are the corrugated containing case built up of sheet steel,

Fig. 139.

and the porcelain terminal bushings with the mode of support for the same.

Fig. 141 represents a circular shell type transformer of the Berry pattern as made by the British Electric Transformer Co., Ltd., who have kindly supplied the information and drawings from which the figure was drawn.

The mode of construction is somewhat special and is of particular interest. The plates forming the middle and bottom portions of the core are first assembled in a suitable mandril, the outer side of the central portion of the core being then covered by a layer of webbing, for mechanical protection of the windings, followed by a layer of mica to serve as insulation between the windings and the core.

Over this is placed one-half of the low tension winding and this, in turn, is covered by further insulation, and an earth shield, if

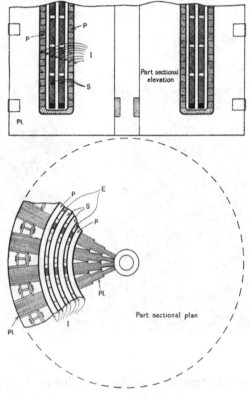

P	Primary winding	I	Insulation
S	Secondary winding	E	Position of earth shields (if any)
	Pl	Stampings	

Fig. 141.

used, is put on. More mica insulation having been placed over the earth shield the sections into which the high tension winding is divided are slipped over the core, ventilation spaces being secured by distance pieces as shown in the drawing. A further earth shield is then put on (if they are used) which is again insulated on both sides by layers of mica, and the winding is completed by adding the remaining portion of the low tension side. This is covered by another layer of mica followed by a layer of webbing and the active portion of the transformer is then ready for completion by the addition of the outside and top core plates. The whole construction is tightly held together by the clamps, binding and distance pieces shown, and it will be realised that a special feature of this design is the very efficient ventilation obtained.

Losses and Efficiencies of Transformers.

The losses occurring in a transformer may be divided into two parts,

(1) Those occurring in the core, and

(2) Those occurring in the copper windings.

Core losses. These include hysteresis and eddy current losses and, as the core flux has a practically constant maximum value and frequency at all loads, may be taken as being practically independent of the load.

Copper losses. These are due to the passage of the current through the resistance of the windings; they will occur both in the primary and secondary and in each case will increase with the square of the current, that is, with the square of the load. In calculating these losses the resistance should be taken for the usual working temperature, allowance being made for the skin effect, if necessary, when massive conductors are used.

Since the core losses are, as mentioned above, practically constant at all loads, it follows that even at no load a small current component in phase with the applied pressure will be taken by the primary winding, and the total no load current of the transformer may be looked upon as being compounded of the magnetising current of the transformer and the small power component due to the core losses, these two components being at right angles to each other as regards phase (see Fig. 142).

The static transformer is one of the most efficient of electrical appliances and in large sizes a full load efficiency as high as 98·5 % may be attained, and even in sizes of 5 to 10 K.W. efficiencies of the order of 96 to 97 % may be expected.   The power efficiency of a transformer at any load is of course the value of the ratio $\dfrac{\text{Output}}{\text{Input}}$; this may also be expressed as

$$\frac{\text{Output}}{\text{Output} + \text{loss}}.$$

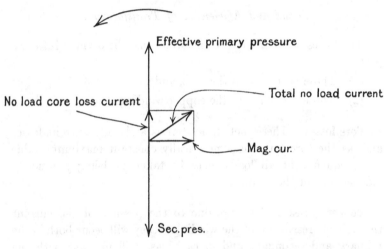

Fig. 142.

The shape of the curve connecting the load with the efficiency of a transformer is influenced very largely by the allocation of the full load losses of the transformer, and it is possible for two trans- formers with equal full load efficiencies to have very different shapes of efficiency curves.   Consider two transformers each with an output of ten K.W. and with an efficiency of 96·1 % at full load and, further, let the full load copper loss of transformer A be 200 watts and that of B be 300 watts; the core loss of transformer A will then be 200 watts and that of B will be 100 watts.   Working

out the efficiencies of the two transformers at several loads we
then obtain the following results:

Load	A				
	Core loss (watts)	Copper loss (watts)	Output (watts)	Input (watts)	Efficiency
No load	200	—	0	200	0·0 %
¼ load	200	12·5	2500	2712·5	92·2 %
½ load	200	50·0	5000	5250·0	95·2 %
¾ load	200	112·5	7500	7812·5	96·0 %
Full load	200	200·0	10000	10400·0	96·1 %
1¼ load	200	312·5	12500	13012·5	96·0 %

Load	B				
	Core loss (watts)	Copper loss (watts)	Output (watts)	Input (watts)	Efficiency
No load	100	—	0	100	0·0 %
¼ load	100	19·0	2500	2619	95·6 %
½ load	100	75·0	5000	5175	96·6 %
¾ load	100	169·0	7500	7769	96·6 %
Full load	100	300·0	10000	10400	96·1 %
1¼ load	100	469·0	12500	13069	95·7 %

In compiling this table the copper loss due to the magnetising
current has been neglected and the losses have been worked out
to the nearest half watt. The efficiency curves are plotted in
Fig. 143.

An inspection of the tables and curves will show that transformer A (*i.e.* the one with the proportionally small full load
copper loss) gives decidedly poorer efficiencies at low loads and
better efficiencies at large loads, while the reverse is of course the
case with transformer B (the one with the proportionally large
full load copper loss).

In passing we may note that transformer B would be
particularly suitable for a lighting load, in which case the transformer may have to run for long periods on little or no load, when
the small core loss will be very advantageous. Also note that
though the transformers have equal full load efficiencies they have
not equal maximum efficiencies, transformer B being superior in
this respect; the maximum efficiency occurs at the load for
which the core and copper losses are equal and for transformer B

this is at about 58 % of full load when the efficiency is seen to be 96·65 %.

When considering the efficiency of electrical machines it is most commonly the efficiency at full load that is dealt with but in many cases, particularly perhaps in transformer working, what is known as the all day efficiency is a better criterion of the performance of the machine.

The all day efficiency may be defined as

$$\frac{\text{Kilowatt hours output from the transformer in 24 hours}}{\text{Kilowatt hours input to transformer in the same time}},$$

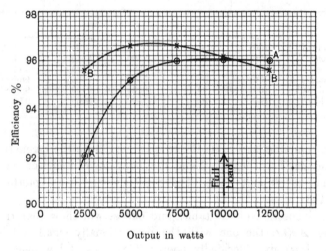

Fig. 143.

and its value depends not only upon the power efficiency of the transformer at any particular load but upon its power efficiency at all loads and also upon the way in which the load upon the transformer is distributed during the 24 hours. Let us suppose that the two transformers already considered are employed on a lighting load which, for the sake of simplicity, we will consider to be constituted of five hours at full load, five hours at half load and the remaining 14 hours at little or no load (that is to say during this period the copper loss will be negligible but the core loss will still go on). A short calculation will then give the following table:

	Transformer A		Transformer B	
Output in 24 hours	75·0	K.W.H.	75·0	K.W.H
Core loss in 24 hours	4·8	,,	2·4	,,
Copper loss				
5 hours at full load	1·0	,,	1·5	,,
5 hours at half load	·25	,,	·375	,,
Input in 24 hours	81·05	,,	79·275	,,
All day efficiency	92·5 %		94·6 %	

It is seen that transformer B has a much superior all day efficiency and this is due to the better efficiency at small loads which in turn is due to the comparatively small core loss. If the transformers were used for a service (as perhaps on power work) such that when in use at all they were working at or near full load with occasional periods of over load, transformer A might give the better all day efficiency due to the greater copper losses occurring in B at over load.

Pressure drop and regulation of transformers. If the pressure applied to the primary of a transformer be kept constant, the pressure at the terminals of the secondary will, in general, fall with increase of load. Important sources of this drop in pressure are the primary and secondary resistances. Considering the former first, we see that a portion of the applied pressure will, in accordance with Ohm's law, be utilised in driving the current through the primary resistance and the magnitude of this component will increase with increase of load.

Since the total pressure applied to the primary is constant at all loads, the portion of it which is available to balance the back primary pressure will decrease with increase of load and therefore the induced pressure, both in the primary and secondary windings, will decrease with increase of load.

Again, when the secondary pressure is sending a current, a portion of the generated secondary pressure will be utilised in driving the secondary current through the secondary resistance, and the magnitude of this component will increase with increase of load and thus there will be less pressure available at the secondary terminals the greater the load.

It should be understood that the pressure drops due to resistance in the primary and secondary windings are not necessarily subtracted arithmetically from the applied and generated pressures as the case may be, they will in all cases be subtracted vectorally; this point is dealt with more exactly later but in any case it does not interfere with the conclusions arrived at above.

Another important source of drop in pressure with increase of load is magnetic leakage. Hitherto we have supposed that all the flux generated in a transformer core links with both windings and hence is useful flux, but in practice, no matter to what extent the primary and secondary windings are interleaved, the primary current sets up a flux (known as the primary leakage flux) which does not link with the secondary; similarly the secondary current sets up a flux (known as the secondary leakage flux) which does not link with the primary. At first sight it may not be at all

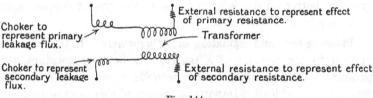

Fig. 144.

clear why these leakage fluxes should cause a drop in pressure but a brief consideration will show that the leakage fluxes cause the windings to behave to some extent as choking coils, and in fact the phenomena that result from leakage are precisely the same as those which would occur if we imagined an actual leaky transformer to be substituted by an ideal transformer provided with external choking coils (of small reactance) placed in series with the windings as shown in Fig. 144. The choking coil in series with either winding produces a small flux which does not link with the other winding and therefore very well represents the leakage which actually occurs in a commercial transformer.

When current flows through these chokers (no matter whether they are internal, as in the actual transformer, or external, as in the imaginary case) a pressure drop will exist across them just as in the case of any other inductive circuit and this drop must be subtracted vectorally from the applied or generated pressure, according as to whether we are considering the primary or secondary winding, in order to obtain the effective pressure in each case.

The precise way in which these resistance and reactance pressure drops influence the resultant pressure drop will depend upon the nature of the load, and in looking into this matter it will

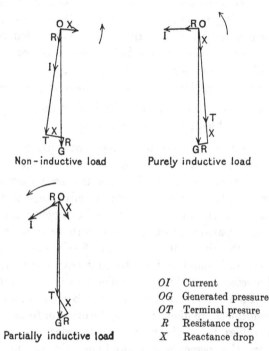

Non–inductive load   Purely inductive load

Partially inductive load

OI	Current
OG	Generated pressure
OT	Terminal presure
R	Resistance drop
X	Reactance drop

Fig. 145.

be convenient for us to imagine the actual resistances and leakage reactances of the two windings to be replaced by a single resistance and a single reactance, situated in the secondary circuit, and of such magnitudes as to give the same drop as would actually occur under working conditions. No matter what may be the phase of the secondary current relative to the secondary pressure the resistance drop will always be in phase with the current, and the leakage reactance drop will always be a quarter of a period in front of the current; remembering these facts the vector diagrams for a non-inductive, a purely inductive, and a partially inductive load given in Fig. 145 are readily arrived at.

In each of these diagrams the resistance and reactance pressure drops are shown in their proper phase relative to the current and the terminal pressure is obtained in each case by subtracting the pressure drops from the generated pressure. It will be seen that on non-inductive loads resistance is likely to be a more important factor in producing terminal pressure drop than reactance since the resistance drop is more nearly directly subtracted from the generated pressure, whereas on inductive loads reactance (or leakage) is likely to be more important.

The percentage regulation of a transformer may be defined as

$$\frac{\text{Rise in secondary pressure from full load to no load}}{\text{Full load secondary pressure}} \times 100,$$

the applied pressure and frequency being kept constant.

The advantage of having good or close regulation on a transformer is that as the load varies the secondary pressure will be more nearly constant, and the disadvantage is that the effects of a short circuit are likely to be much more severe. Resistance and leakage reactance will both tend to limit the magnitudes of the short circuit currents flowing in the primary and secondary and therefore tend to limit both the thermal and mechanical stresses occurring in the windings under short circuit conditions. On unity power factor the regulations commonly met with in transformers are likely to be of the order of from 1 to 3 %, the smaller values being those of large transformers. At lower power factors the percentage regulations will be, in general, somewhat higher.

The Auto-Transformer.

It is possible to arrange a device which, while working on ordinary principles, has but one winding, the whole of which acts as the primary and a part as the secondary or *vice versâ*. Consider the diagram shown in Fig. 146, in this case a certain pressure is applied to the whole of the winding and in each turn a certain induced pressure, opposite in phase to the applied pressure, will

be produced. The induced pressures in the part of the winding
which is common to both primary and secondary circuits serve
two purposes, in the first place they help to oppose the applied
pressure in the primary circuit, and in the second place they may
at the same time cause current to flow in the secondary circuit.
It will be realised that the turns common to both windings will
be traversed by both the primary and secondary currents and the
actual current in these turns will be the vectoral sum of the two
and since, apart from the magnetising component of the primary
current, the two currents are in opposite phases, a considerable
neutralisation will result. This indicates the possibility of a
considerable saving of copper in the auto-transformer as compared
with the ordinary transformer and this is the chief reason for the

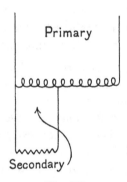

Fig. 146.

employment of the former device. All the turns are wound on
a common core as in an ordinary transformer.

The amount of copper which can be saved depends very much
upon the pressure ratio of the device and in the comparison
effected in the following example a constant current density is
taken in all conductors, the copper required for the magnetising
current is neglected (it is quite unimportant at or near full load),
and as a convenient unit the amount of copper required for one
turn capable of carrying 5 amperes is adopted.

Case I. Step down transformer with a pressure ratio of 1 : 2
and with a secondary current of 10 amperes. Primary turns = 100.
In this case the secondary current will be tapped off from

50 turns and the primary current will be 5 amperes if the magnetising current is neglected. The current flowing in the turns common to both windings will be the resultant of 10 and 5 amperes in phase opposition and will of course be 5 amperes.

Copper required in ordinary transformer.

Primary winding	...	...	...	100 units
Secondary winding	...	...	...	100 ,,
Total ...	...	...	...	...	200 ,,

Copper required in auto-transformer.

Common part of the winding ...	...	50 units
Remaining part of the winding	...	50 ,,
Total ...	...	...	...	...	100 ,,

The saving in copper obtained by the adoption of the auto-transformer is seen to be 50 % in this instance.

Case II. Step down transformer with a pressure ratio of 1 : 4 and with a secondary current of 10 amperes. Primary turns 100.

The secondary winding will now comprise 25 turns and the primary current will be 2·5 amperes, the magnetising current again being neglected. The current flowing in the turns common to both windings will be the resultant of 10 and 2·5 amperes which are in phase opposition and this amounts to 7·5 amperes.

Copper required in ordinary transformer.

Primary winding	...	...	...	50 units
Secondary winding	...	...	...	50 ,,
Total ...	...	...	...	...	100 ,,

Copper required in auto-transformer.

Common part of the winding	...	37·5 units
Remaining part of the winding	...	37·5 ,,
Total ...	...	...	...	...	75 ,,

The saving in the amount of copper is now 25 %.

Generalising, we see that the auto-transformer permits of the greater percentage saving of copper the nearer the ratio of transformation approaches unity.

In practice, it is only desirable to use an auto-transformer when the pressures of the primary and secondary are not very dissimilar or when they are both low, since the risk of the high pressure appearing in the low pressure winding is greater than in the ordinary transformer. Thus, to take a concrete instance, suppose

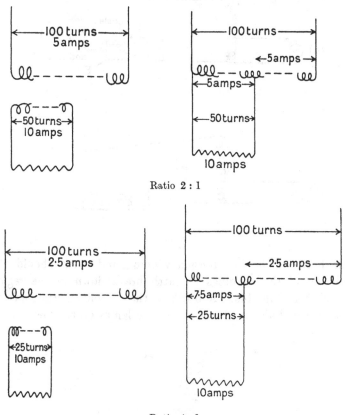

Ratio 2 : 1

Ratio 4 : 1

Fig. 147.

that an auto-transformer is used for step up purposes in connection with the production of a high pressure for cable testing purposes as shown in Fig. 148 (a) and (b); if the connections are correctly made as in (a) no harm will result, but if, in error, the connections are made as in (b) we should have the full pressure of the high

tension side of the transformer occurring between the winding of
the low tension alternator and earth with a consequent breakdown
of the slot insulation of the alternator.

Again, were auto-transformers used for step down purposes in
connection with lighting schemes, a considerable pressure would

(a)

Cable under test

Tank filled
with water
and earthed

E

(b)

E

Fig. 148.

occur between the low tension winding and earth should the
insulation of the windings to earth break down in the region
marked E in Fig. 149. A break in the common portion of the
winding would also be disastrous to the lamps on the low tension
side.

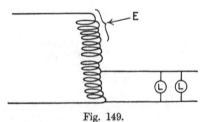

E

Fig. 149.

In practice the auto-transformer is used to a considerable
extent for stepping down the line pressure in starting induction
motors (in this connection it is often termed a compensator), and
also in running a single arc lamp off say a 200 volt alternating

current main (it is then often termed an economy coil). In the latter connection it has great advantages over a choking coil, which can be used for the same purpose, since it allows of a less current at a higher power factor being taken from the mains. Recently it has also been used to a large extent for running low pressure metallic filament lamps off 200 to 250 volt mains.

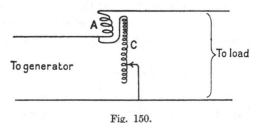

A

C

To generator

To load

Fig. 150.

Safe and efficient use of an auto-transformer may also be made for boosting purposes on high or low pressure circuits as indicated in Fig. 150. In this case the portion of the winding marked C only carries a small current and therefore does not involve the use of large switch contacts (which must be designed so as not to short circuit any turns or break the circuit when switching from

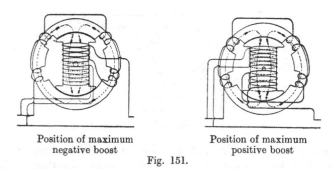

Position of maximum
negative boost

Position of maximum
positive boost

Fig. 151.

one contact to another as in accumulator switches). The less turns there are in the part C the greater will be the boost produced.

If desired, switch contacts may be dispensed with altogether and at the same time a perfectly continuous graduation of the boost obtained, by adopting the connections and arrangements indicated in Fig. 151 (kindly furnished by Messrs Switchgear and Cowans, Ltd.) which represents the Cowan Still regulator. In this case the winding which corresponds to the winding A in the previous figure is wound on the outer fixed laminated cylindrical core, and the winding corresponding to C is mounted upon the central movable core. The position of C relative to A may be altered by means of a worm wheel gearing, and it is clear that placing C in different positions will result in a varying degree of boost in A. It is usual to arrange for the maximum boost to be one-half of the maximum pressure variation required and this boost may be used either to assist or oppose the main pressure.

Phase Transformation.

A problem which occasionally arises in connection with transformers is that of the production of alternating currents having a given number of phases when the supply pressure has a different number of phases. In certain cases this is obviously a very simple matter as, for instance, when single phase current is required and the supply is two or three phase.

The problem can also be readily solved generally if the use of rotating machinery is permissible. What is almost a general solution can be effected with the help of an appliance illustrated

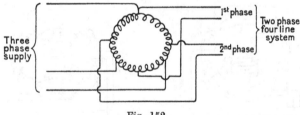

Fig. 152.

diagrammatically in Fig. 152, in which a winding, shown of the ring pattern for the sake of convenience, is arranged on an iron core. If a supply system of any number of phases (except of course single phase) is connected to suitable points of the winding a rotating magnetic field will be produced, and if secondary tappings are taken off from the correct points a system of multiphase currents having any required number of phases may be obtained. The most common cases of phase transformation required in practice are the transformation from two to three phase or *vice versâ*, and the transformation from three to six or twelve phases, and these can be effected with the help of ordinary transformers.

The former is perhaps the most interesting case and can be readily effected if three single phase or a three phase transformer can be made use of. Let us suppose that we are converting from three to two phases and that the transformers have equal turns on the three phase side. The connections are as shown in the diagram in Fig. 153 from which it will be seen that one phase of the two phase current is obtained directly from the action of one limb of the three phase side, while the second phase of the two phase side is obtained from a combination of the actions of the other two limbs of the three phase side. Now the pressures in the two phases of the two phase side must be equal to each other

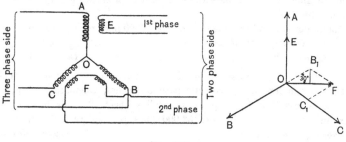

Fig. 153.

and, supposing that the number of turns in winding E is 1000, let us determine the number of turns in each of the windings which go to make up the phase F. In the vector diagram the two pressures whose sum gives the phase pressure F are shown as OC_1 and OB_1 and, assuming that equal pressures are induced in all individual turns, we have

$$\frac{\frac{1}{2}OF}{OB_1} = \cos 30°, \text{ whence } OB_1 = \frac{\frac{1}{2}OF}{\cos 30°} = \frac{500}{\frac{\sqrt{3}}{2}} = 577 \text{ turns.}$$

That is to say for every 1000 turns we have in E we shall need practically 577 turns in each of the windings which together go to make the phase F. A little thought will show a method whereby the same phase transformation may be effected by the use of two transformers.

EXAMPLES

1. A transformer is required to step down from 6500 to 460 volts. If there are 5000 turns on the primary winding how many must be used in the secondary if the effects of resistance and magnetic leakage are neglected?

Answer. 354 turns.

2. Draw to scale a vector diagram for the transformer to which the following numbers apply:

Primary pressure	1000 volts.
Secondary pressure	200 volts.
Magnetising current	1·5 amperes.
Core loss current	1 ampere.
Secondary current	40 amperes.

Phase difference between secondary current and secondary pressure = 30°.
Pressure drops due to resistance and leakage reactance may be neglected.

3. How many turns must be placed on the secondary of a transformer to which the following particulars apply in order to produce a generated pressure of 200 volts?

Maximum value of flux density in core = 5000 lines per square cm.
Area of cross-section of core 60 square cms.
Frequency = 50 cycles per second. *Answer.* 300 turns.

4. If the area of cross-section of a transformer core is 10 square inches and the mean length of magnetic path 50 inches, calculate the approximate core loss if it is composed of Stalloy sheets 20 mils in thickness. Maximum flux density in core 52,000 lines per square inch; frequency 50 cycles per second. *Answer.* 80 watts.

5. Repeat the above example if the core is composed of ordinary transformer iron. *Answer.* 151 watts.

6. Two 5 K.W. transformers have the following allocation of full load losses:

	A	B
Core loss	50 watts	150 watts
Full load copper loss	150 ,,	50 ,,

Plot curves connecting efficiency and load for each transformer.

7. State the loads in K.W. for which each of the above transformers will have its maximum efficiency. *Answer.* A 2·9 K.W.; B 8·7 K.W. (approx.).

8. Calculate the all-day efficiency of each of the above transformers if they are run on a lighting load constituted as follows:

Full load	6 hours per day.
Half load	4 ,, ,,
Light load	14 ,, ,,

Answer. 94·7 %; 91 %.

9. Repeat the last example for a power load constituted of:

1½ load ...	...	...	...	2 hours per day.
Full load ...	...	...	...	8 ,, ,,
Switched off	...	...	...	14 ,, ,,

Answer. 95·9 %; 96·2 %.

10. If the full load secondary pressure of a transformer is 200 volts and the equivalent resistance and reactance pressure drops 4 and 10 volts respectively, determine the necessary generated secondary pressure when the load has power factors of 1 and ·8 respectively. Determine also the percentage regulation in the two cases.

Answer. 204·2 volts; 209·3 volts; 2·1 %; 4·6 %.

11. In the case of a pressure ratio of 1 : 3 determine the percentage of copper saved by adopting the auto-transformer in preference to the ordinary transformer. Use equal current densities in all cases. *Answer.* 33·3 %.

12. Draw a diagram showing the method of connection of two single phase transformers used for the purpose of transforming from two to three phase.

CHAPTER IX

INDUCTION MOTORS

The motor most frequently met with on alternating current circuits is undoubtedly the induction motor, and the rapid advance which this type (particularly for polyphase circuits) has made during the last thirty years has been one of the most striking features of the progress of electrical engineering. Two separate and independent discoveries have brought this about; the first, due to Arago, dates back to the year 1824, and the second, due to Ferraris, to about 1885.

Arago discovered that if a conductor was placed in a moving magnetic field eddy currents were produced within it, and the interaction between the eddy currents and the magnetic field were such as to cause a mechanical force to be exerted tending to drag the conductor in the same direction as the motion of the magnetic field. There will be of course the usual reaction, in this case tending to stop the motion of the magnetic field. The form of apparatus originally used was very simple and consisted of a copper disc, arranged so as to rotate about a vertical axis, suspended above a permanent horse shoe magnet also capable of rotation about the same vertical axis. A glass plate was placed between the poles of the magnet and the disc to prevent the possibility of the motion of the disc being influenced by the air currents set up by the motion of the poles of the magnet. On rotating the magnet, the lines due to the magnet are also caused to rotate (we have in fact a crude form of rotating magnetic field) and these, by cutting the disc, set up eddy currents in the copper having the directions and paths indicated in Fig. 154.

The bands of current act as small solenoids and have a definite polarity both on the upper and lower sides of the disc. In the figure the poles of the rotating magnet are supposed to be under

the disc and the polarities of the solenoids shown are those on
the under side of the disc which are, of course, nearer to the poles
of the magnet. The band of current (and the disc with it) in
front of the moving pole is pushed round, and the band of current
behind the pole is pulled round, the resultant motion of the disc
being in the same direction as the motion of the poles. Of course
the polarities on the upper side of the disc would tend to give
a motion in the reverse direction, but, since these poles will be
farther away from the poles of the magnet, their effect will be
more than neutralised by the forward rotating effect of the
solenoidal poles under the disc.

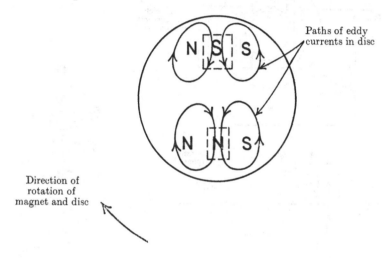

Paths of eddy
currents in disc

Direction of
rotation of
magnet and disc

Fig. 154.

It is important to notice even at this early stage that the disc
cannot possibly move so fast as the magnet, since if this occurred
there would be no relative motion between the disc and the
magnetic lines, and so neither induced pressure, current or
mechanical force would be produced. A certain amount of slip
between the moving magnetic field and the copper disc is essential
to the correct action of the motor. If the disc is cut in such a
way as to minimise the eddy currents the force tending to drive
the disc will be diminished and if the eddy currents are absolutely
stopped no mechanical force whatever will be developed.

The principle in operation in the above experiment is precisely that made use of in the modern induction motor, but the means adopted for producing the rotating magnetic field are of course quite inapplicable to a commercial machine, and the commercial application of the induction motor was only rendered possible by the second of the discoveries mentioned above which indicated a method of obtaining a rotating magnetic field, without making use of any material moving parts, by using polyphase currents.

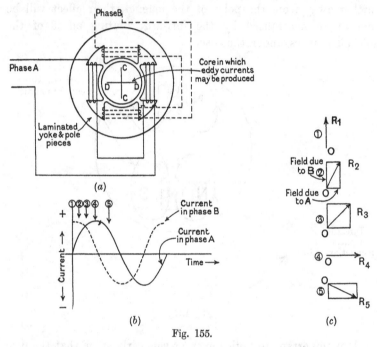

Fig. 155.

Imagine a laminated yoke provided with four inwardly projecting radial poles as in Fig. 155, and let two opposite poles be wound with a winding connected to the first phase, and the other two opposite poles be wound with a winding connected to the second phase, of a two phase alternating current supply. The two currents will differ in phase by 90° and the first current will tend to produce an alternating field along a horizontal line while the second current will tend to produce an alternating field along a vertical line.

At the instant marked (1) the current in phase *A* will be zero while the current in phase *B* will be a maximum and hence in the core a resultant magnetic field will be produced along the line *CC*. A little later the current in phase *B* will have fallen somewhat while the current in phase *A* will have risen to some small value, the field in the core will now be compounded of a strong component along the line *CC* and a small component along the line *DD*, the resultant field being shown by OR_2 in the diagram. At the instant (3) the component due to phase *B* will have fallen to a still lower value while that due to *A* will have risen, the resultant now being shown by OR_3. At the instant (4) the component due to *B* will have fallen to zero while the component due to *A* will be a maximum, the resultant being represented by OR_4. Finally, at the instant marked (5), the component due to *B* has again a small value, but this time in the opposite direction, while the component due to *A* has begun to fall, the resultant being represented by OR_5.

Thus we see that as the two phase currents pass through their cycles we have produced in the gap and core a rotating magnetic field having precisely the same characteristics as the rotating fields of Arago, but obtained in a far more convenient manner, and if we place in this rotating field a body in which eddy currents may be produced we shall obtain mechanical motion of the body just in the same manner as in the original experiment.

It is worth noting that the magnetic circuit shown really gives a two pole magnetic field and not a four pole field as might at first sight be supposed. Though the type of magnetic field with the salient poles is very convenient for an elementary explanation it is not used in practice except for supply meters and other small devices working on this principle. All machines of this class intended to develop any appreciable power are provided with what is practically a smooth core in which tunnels or nearly closed slots are placed to receive the windings, which are almost invariably of the drum type though a ring winding might be used if desired. A simple piece of apparatus intended to illustrate the principle of the induction motor is perhaps worthy of mention, it consists of a smooth ring of laminated iron over which is placed a ring winding suitable for two or three phase currents. The two parts into which each phase of the winding is divided are arranged to produce consequent poles in the iron ring with the result that a field is produced in the interior of the ring as shown (for one phase only) in Fig. 156. In the interior of the ring an egg-shaped mass of aluminium or copper is placed on a suitable stand, the interior of the metal often being hollow and filled with iron filings in order to improve the path of the magnetic lines. When current is switched on to the windings a rotating field is produced in the usual manner and the egg commences to spin and if originally placed on its side will often raise itself upon end—a most striking experiment.

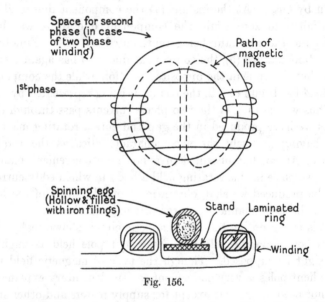

Fig. 156.

From what has been stated the reader will realise that any polyphase induction motor comprises two essential parts:

(1) The part containing the winding adapted to receive the line currents. This part is usually stationary and is known as the *Stator*, its function is to give rise to the rotating magnetic field.

(2) The part in which the moving magnetic field is to produce the induced currents, the interaction between which and the moving field produces the mechanical forces. It is commonly known as the *Rotor*.

In the above investigation the reader may perhaps have gathered the impression that the flux in the gap is produced by the magnetic effect of the primary winding alone, but it is important to be quite clear that this is not the case. The actions taking place are similar to those occurring in the transformer (and in thinking of the currents and fluxes in an induction motor it is always desirable to keep transformer principles in mind). The useful flux of the induction motor (*i.e.* that linking with both windings) represents the resultant effect of the stator and rotor currents, just as in a transformer the useful flux represents the resultant action of the primary and secondary currents. If the rotor current was zero (as might happen with a wound rotor which is open circuited and therefore producing no torque) the core and gap flux would be produced by the stator magnetising current alone, but, when any rotor current flows, an additional current flows in the primary winding and the magnetic effects of the rotor current and the additional primary current will, so far as the useful flux is concerned, neutralise each other thus leaving, as in a transformer, the magnetising current free to produce what is practically a constant flux at all loads.

In addition to the useful flux, the stator current (as a whole and not merely the magnetising component) will produce a primary leakage flux, and the rotor current will produce a secondary leakage flux. These leakage fluxes have considerable effect on the power factor and performance of the motor and in practice every care is taken to keep them as low as possible.

General arrangements of the commercial induction motor. A half sectional elevation of a modern type of induction motor is shown in Fig. 157 and a photograph of a very similar completed machine in Fig. 158. It will be seen that the machine shown in the photograph is represented with bush bearings while the one shown in the drawing has ball bearings. Both views have been kindly supplied by the Phoenix Dynamo Manufacturing Co. of Bradford.

The stator. This comprises a cast iron housing H which serves as a support for the actual core SC. The core is usually composed of stampings of alloyed iron some 20 mils in thickness. The inner

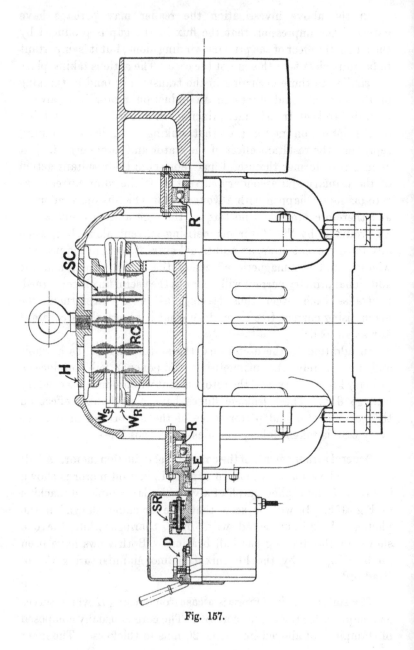

Fig. 157.

face of the laminations is occupied with the slots arranged to
receive the windings and the exact shape of these slots is a matter
of some importance. A smooth core winding is of course out of
the question, since it would be poor from the mechanical point of
view, and would also involve a considerable length of gap with
a consequently large magnetising current. The choice then lies
between slots and tunnels; the former interfere with the uniform
progression of flux along the stator face*, and the latter tend to

Fig. 158.

give rise to too much leakage since a convenient magnetic path
is provided round the conductor which does not allow of the
lines passing through it linking with the rotor windings.

In practice a compromise is made between the conflicting conditions outlined above and a nearly closed slot is employed; this,
while obviating undue heating of the teeth, at the same time puts
a series of non-magnetic gaps in the path which the leakage flux
mentioned above tends to follow and so very materially reduces

* If the flux does not progress at a uniform rate it is liable to give rise to
excessive heating in the teeth.

its amount. All the remarks made above concerning stator slots apply with equal force to the rotor slots. The stator windings made use of are similar to those employed for the armature windings of alternators, namely, two layer wave windings for low pressure machines and coil windings for high pressure machines. In Fig. 157 the stator winding is shown at W_S.

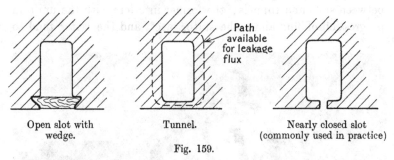

Open slot with wedge.   Tunnel.   Nearly closed slot (commonly used in practice)

Fig. 159.

The rotor. The rotor must provide a good path for the magnetic lines in a plane at right angles to the shaft, and a good path for the current in lines parallel to the shaft.

It is necessary therefore that the rotor core be composed of iron and also laminated in a plane at right angles to the shaft.

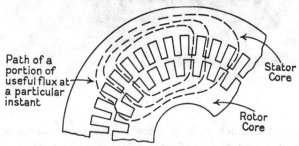

Diagram of flux in rotor and stator core and teeth.

Fig. 160.

The laminations of the rotor core need not be made very thin (though commonly they are made, for convenience, out of the same sheets of material that are used for the stator stampings) and in fact lamination is not absolutely necessary since, when the rotor is running at full speed, the slip, and consequently the frequency of the alternations of the rotor flux, is quite low and so will not cause any considerable heating. A good path for the rotor current is provided by a winding W_R placed in slots in the rotor core. The simplest type of winding is the so-called squirrel cage rotor; in this case a single stout bar is placed in each slot, the whole of the bars being short circuited at each end by means of stout copper end rings or other similar connections. Since the pressure induced in such a winding is very low but little insulation is needed between the rotor bars and the core. In the case of large squirrel cage rotors it is desirable that the end rings be riveted and sweated to the end of each rotor bar, while in small rotors of this type an efficient end connection may be secured by binding the ends of the rotor bars with a tinned copper lapping having a considerable number of turns as indicated in Fig. 161, the whole being well sweated together to secure good electrical conductivity. The end rings may also be cast on to the rotor bars.

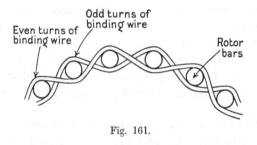

Fig. 161.

Other rotors, chiefly for moderate and large sized machines, are provided with a regular winding, very commonly of the bar wave wound type, starred internally, the free ends being brought out to slip rings, as in Fig. 157, for convenience in starting.

It is important to note that a wound rotor must be arranged for the same number of poles as the stator, though it may be wound for a different number of phases; thus a two phase stator may be provided with a three phase rotor. In a squirrel cage rotor the number of poles is automatically settled by the number of poles of the stator, and one rotor of this type may be used in stators having different numbers of poles without any change of rotor arrangements, a fact taken advantage of in producing multispeed induction motors.

In Fig. 157 the rotor is indicated as being provided with a
three phase winding, the three free ends being brought out to
slip rings *SR* to which, by means of brushes, resistance may be
connected for starting purposes. This motor is also provided
with a short circuiting gear *D* for the rings, and a brush lifting
device in order that resistance and friction losses may be minimised
when running at full speed.

It will be noticed that not only are ball bearings *R* provided
for taking the radial load, but, in addition, small bearings *E* are
provided for taking any end thrust which may develop.

Air gap in induction motors. It is very important that the
air gap in induction motors be as small as is consistent with
satisfactory mechanical operation. A large air gap necessitates
a large magnetising current and causes increased magnetic leakage,
and these, in turn, give rise to a low power factor which it is the
aim of the designer to avoid. In small motors the gap may have
a length of the order of ·02 inch increasing in large motors to
·06 inch. It is necessary to take great care in centring the rotor
in this gap otherwise undue bending moments may develop on the
shaft giving rise to chafing between the rotor and stator cores,
particularly when starting up with squirrel cage rotors*.

Production of torque in the polyphase induction motor. The
forces tending to cause the rotor to revolve may very conveniently
be looked upon as being produced by the current in the rotor
conductors acting in conjunction with the magnetic field produced
in the gap. The latter is due to the resultant effect of the rotor
and stator currents and may be looked upon as having an
approximately constant value at all loads so long as the applied
pressure is kept constant. The torque produced in an induction
motor thus depends upon three factors:

(1) The strength of the field in the gap.

(2) The magnitude of the rotor current.

(3) The phase relation between the resultant field and the
rotor current.

---

* Increased core losses may also result from decentralisation of the rotor, see
paper by Messrs Smith and Johnson, *J. I. E. E.*, Vol. 48, p. 546.

This last point is of particular importance and needs further explanation; at any given instant the variation of flux round the gap is roughly sinusoidal and the distribution of rotor current follows approximately the same law, but the position of maximum field does not necessarily coincide with the position of maximum current. Thus in Fig. 162, which represents a two pole case, the region of maximum flux occurs at *A* and in this same region we shall also have the maximum values of rotor pressure (induced by

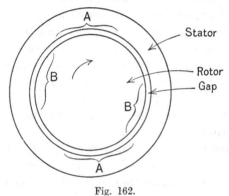

Fig. 162.

the slip of the rotor conductors relative to the rotating field), but if the rotor current lags considerably behind the rotor pressure the bands of rotor current will attain their maximum values in the region marked *B*. Thus we may have a strong field and considerable rotor current and yet have but a poor torque owing to the bad phase relation between the two. We see that the bands of considerable rotor current occur in the regions of comparatively weak gap field.

The Starting of Polyphase Induction Motors.

To successfully start an induction motor it is necessary to produce as large a starting torque as possible with the minimum of line disturbance, that is without taking an unduly large current from the line and also without taking it at a poor power factor. A method of starting a motor may be regarded as very good if it is such as will give full load torque at starting without exceeding full load current. Numerous methods have been devised for the purpose of starting induction motors and the most important are briefly described below together with an account of their advantages and disadvantages.

(1) **Stator switched directly on to the mains when a squirrel cage rotor is used.** In this case it will be seen that immediately the main switch is closed the rotating field will commence revolving at full speed while the rotor is momentarily at rest and afterwards moves but slowly for an appreciable period. The frequency of the induced pressures and currents in the rotor will, at the moment of starting, be high, falling off as the rotor speeds up. Now, since the resistance of this type of rotor is very low, the ratio of reactance to resistance in the rotor circuit will be high at the moment of starting with the result that the rotor current will lag considerably behind the rotor pressure and we have already seen that this is a state of affairs that will produce but little torque. Again, since the rotor acts as a short circuited secondary to the stator, considerable currents at a low power factor will be taken from the line. The starting torque per ampere of current taken from the line will be very low compared with the torque which will be obtained per ampere under running conditions and this method is only suitable for small motors as, for instance, loom motors used for individual driving. The line disturbance produced is likely to be relatively considerable.

(2) **Use of a star delta starting switch with a squirrel cage rotor.** When this method is used the three phases of the stator are not inter-connected in the machine but are brought separately to the starting switch which is provided with contacts so that the phases may be starred for starting (thus putting a greater impedance across the lines) and meshed for running. A little thought will show that we shall again obtain a poor starting torque per ampere but that there will be less line disturbance than in the former case. The actual connections of such a starter are shown in Fig. 163 in which the black circles indicate the fixed contacts of the starter which are connected to the machine and line leads as shown in the diagram. The handle has three successive positions—off, star (for starting), and mesh (for running), thus ensuring that the star position must be passed before the full running position is reached.

In the off position the fixed contacts are not bridged across in any way, but in the star position they are bridged as shown by the dotted lines and in the mesh position as shown by the chain lines.

(3) **Use of a transformer or auto-transformer in the stator circuit for motors having a squirrel cage rotor.** It will be clear that, for the same reasons as in the two previous cases, this will not be a good method for the production of considerable starting torque per ampere of line current, and the chief advantage of this method of starting induction motors is that the line disturbance is not so great as when the above mentioned methods are employed. Transformers for this purpose are usually provided with several tappings so that the most suitable pressure for any particular

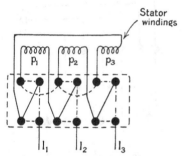

Star-Mesh starter for induction motor.

Fig. 163.

case may be employed, but the best tapping, having been found by trial for any particular motor, is permanently connected up. The tapping used should be that which gives the least pressure on the motor that will ensure satisfactory starting for the conditions under which the motor is working. Since the pressure applied to the motor for starting purposes is less than the line pressure, the current taken by the motor itself will be considerably reduced, and, owing to the usual action of the transformer, the line current will be less than that taken by the motor by an amount depending upon the particular tapping in use. When the motor has practically attained its full speed the transformer is cut out and the full line pressure applied to the motor. Fig. 164 shows the diagram of connections of such a starter and is of

particular interest since not only is the auto-transformer principle
made use of but one phase is suppressed entirely, the pressure for
the corresponding phase of the motor being supplied by combining
the pressures of the other two phases as is clearly indicated in the
figure.

The fixed contacts of the starter are again indicated by the
black circles and the starting handle (which has the three successive
positions—off, starting and running, thus ensuring that use is
made of the correct sequence in starting up) carries insulated
bridge pieces.   In the off position no connections are made to the
fixed contacts; in the starting position the fixed contacts are

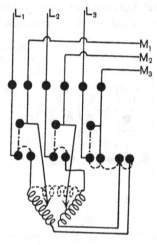

Fig. 164.

bridged as shown by the dotted lines thus bringing the auto-
transformer into action; and, finally, in the running position, the
bridge pieces make contact as shown by the chain lines, thus
applying the full line pressure to the motor and cutting out the
transformer windings altogether. It should be noted that in
such auto-transformers very high current densities may be used
in the copper (especially if the device is oil immersed) because
they are only used for a short length of time and in the interval
between successive starts there is considerable time available for
cooling down.

(4) **Use of resistance in stator circuits for motors having squirrel cage rotors.** A type of starter sometimes used in connection with squirrel cage motors which are not required to produce any considerable starting torque is one which inserts resistance in each phase of the stator circuit, the resistance being gradually cut out as the motor speeds up. The connections and general arrangements of such a starter are shown in Fig. 165, for which and for information concerning the same the writer is indebted to Mr G. Ellison of Birmingham. An inspection of the diagram shows that

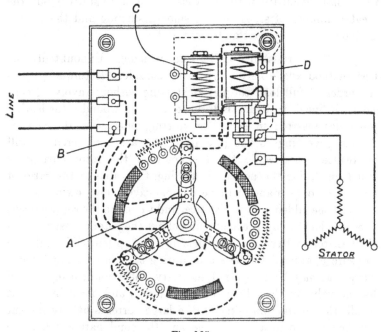

Fig. 165.

the resistance is cut out by an arm having three prongs, each of which carries an insulated brush which serves to make contact between a brass sector A, connected to one of the line wires, and the corresponding set of contact studs B. This arm is pulled to the off position (in which event the pressure is completely cut off from the motor thus dispensing with the necessity of a separate main switch) by a spring, but when the handle is in the full on position it is held against the pull of the spring by a catch (not indicated in the diagram) pulled up by the no volts release C. The over load release works for one line only and operates, when the load becomes excessive, by pulling up its plunger which in turn knocks off the catch of the no volts release thus allowing the spring to return the handle to the off position. These starters are usually adjusted to allow about $2\frac{1}{2}$ times the full load current to pass on the first notch and it is stated that under these conditions the motor develops from $\frac{1}{3}$ to $\frac{1}{2}$ of the full load torque. Of course the current taken from the line is equal to the current taken by the motor (compare with the case of the auto-starter) but the greater number of steps ensures smooth starting and the starter is made in a convenient and compact form.

(5) **Use of resistance in the rotor winding.** Undoubtedly the best method available for producing considerable torque (say of the order of full load torque) at starting without involving considerable line disturbance, is to make use of a wound rotor having resistance inserted in series for starting purposes. If the conditions governing torque (see page 234) are considered, it will be realised that while the frequency of the rotor current is still high during the period of starting, the value of the ratio of reactance to resistance will be comparatively low, owing to the effect of the added rotor resistance, and therefore the rotor current does not lag to any great extent behind the rotor pressure. We shall therefore have a strong field in the gap (since the full line pressure is applied to the stator during starting), as much rotor current as may be considered necessary (the amount being settled by the value of the added rotor resistance) and, most important of all, the conductors carrying the large currents at any instant will be situated in a strong field and the combination will cause a very good starting torque to be developed.

As a matter of fact full load starting torque can be secured with little more than full load rotor current (a much less value than would be required to attain the same end if other methods of starting were used); further, it follows, from ordinary transformer principles, that if the rotor current is moderate the stator current will also be moderate, and since the angle of lag in the rotor circuit is small the angle of lag in the stator circuit will also be small and the line disturbance will be a minimum.

The starting resistance used may be of the metallic or of the liquid type, and it should be noted that the resistance of the various contacts, leads to the resistance, etc. will be objectionable when running at full speed owing to the loss of power and increased slip which they will occasion, but, if desired, this trouble can be obviated by short circuiting the slip rings internally and lifting the brushes when full speed is reached.

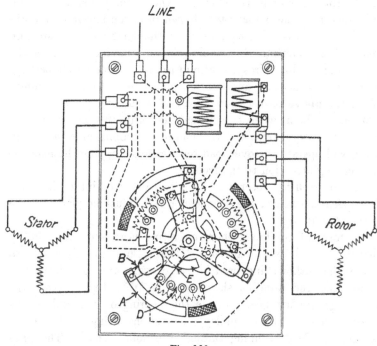

Fig. 166.

The complete connections of a rotor resistance type of starter arranged to control both the rotor and stator circuits are shown in Fig. 166 which has been placed at the disposal of the writer by the maker, Mr C. Ellison. The line wires are brought to the contacts marked A and the moving arm carries three insulated bridge pieces B, each of which bridges across from one of the brass sectors A to the smaller sectors C to which the stator leads are connected; when the handle is in the off position the stator is seen to be entirely disconnected from the line and thus the use

of a separate main switch is obviated.   The three rotor leads are
connected to the studs and resistances marked D, the latter being
gradually short circuited by the arms E which form a neutral
point for the rotor resistance.   The no volts release is connected
across two line wires and the over load release in one of the rotor
leads, the mechanical arrangements of these being as in the
starter last described.   The object of placing the over load release
in the rotor circuit is as follows: if the stator pressure failed on
one line only, due perhaps to a blown fuse, the motor would still
continue to run as a single phase motor, though it would not start up
under these conditions, but would take additional current from the
remaining two line wires and also in the rotor circuit thus perhaps
causing over heating and consequent damage to the windings.
With the above connections, however, if the over load release is
not set too high, when the current ceases in any one stator line
the extra rotor current (if the motor is running at or near full
load) will operate the over load release thus stopping the machine
and calling attention to the trouble.

Of course if the rotor winding is to be shorted internally when
full speed is attained the over load release must be placed in the
stator circuit.

A form of combined short circuiting gear and brush lifting
device as used by the Phoenix Dynamo Manufacturing Co., to
whom the author is indebted for the drawing and for information
concerning the same, is shown in Fig. 167.   In this case the
slip rings are arranged outside the bearing on the end of the
shaft remote from the pulley, connections being made to the
winding by leads running through the hollow shaft.   The rings
are surrounded by a substantial skeleton frame the openings in
which are covered by sheet metal.   The plate P at the back is
rigidly secured to the handle C_2, both being free to revolve over
a limited arc, a catch K being provided to secure them in either
of the limiting positions.   When the handle is turned in the
counter clockwise direction (looking towards the motor) the collar
E turns with it and the portion D, which slides in a slot in the
collar, is forced along towards the motor carrying the whole of
the second collar F with it.   The part S fits on the shaft and
revolves with it owing to the key, but is free to slide along axially,

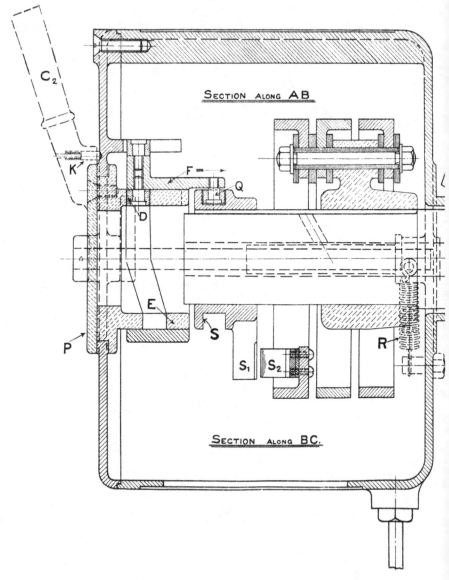

SECTION ALONG AB

SECTION ALONG BC.

Fig. 167. Arrangement of Brush

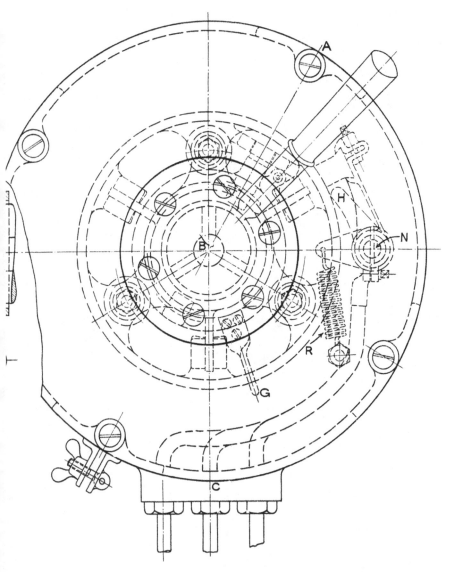

Lifting and Short Circuiting Gear.

thus when F moves towards the motor the part S also moves along being impelled by the pin Q. The rings are therefore shorted by the contacts at S_1 engaging with the contacts at S_2, one of the latter being provided for each ring.

The brush lifting device is not fully indicated in the figure but its action may be followed with the help of the explanation given

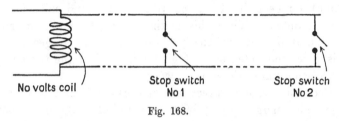

No volts coil   Stop switch No 1   Stop switch No 2

Fig. 168.

below.   The three brushes are mounted on a common spindle but are insulated from each other and from the spindle, the spindle being pulled so as to cause the brushes to make contact with the rings by means of the spring R.   As the portion E is moved round, in addition to shorting the rings, it engages with a projection on the spindle N and turns it round thus lifting the brushes off the

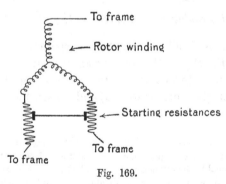

To frame

Rotor winding

Starting resistances

To frame

To frame

Fig. 169.

rings against the pull of the spring.   The handle is kept from flying back by the locking device K.

Sometimes the slip rings are dispensed with and the starting resistances placed inside the rotor or in a part projecting beyond the bearing.   When this method is used the resistance revolves with the rotor and the short circuiting may be carried out automatically by means of a centrifugal device.   One of the best

16—2

arrangements of this type is that due to Mr N. Pensabene-Perez, who employs a device which, by the action of centrifugal force, cuts out the starting resistance step by step thus keeping the starting current practically constant throughout the period of acceleration. Amongst the advantages claimed for this method are simplicity in starting (it is only necessary to close the main switch), no necessity for a no volts release, and suitability for remote control. When this method is used simplicity can be secured, if desired, by the adoption of a two legged starting resistance, connected as in Fig. 169, instead of the three legged resistance usual in three phase work.

Occasionally, as in direct current working, it may be desirable to provide emergency stop buttons or switches at various points adjacent to an induction motor in order that the machine may be stopped without the delay incurred in moving to the main switch. This can be arranged for by running two wires from the no volts coil of the starter, one attached to each end, the switches being connected so as to short circuit the no volts coil as indicated in Fig. 168.

Speed Regulation of Polyphase Induction Motors.

The methods available for this purpose either involve an alteration in the slip of the rotor or an alteration in the speed of the rotating field (which is effected by altering the number of poles for which the motor is wound) and neither method is entirely satisfactory*.

* The synchronous speed of an induction motor, that is the speed of the rotating field, is inversely proportional to the number of poles of the motor and directly proportional to the frequency of the applied pressure, the exact connection between the several quantities being readily obtained as follows, p being the number of poles for which the motor is wound.

$$\text{Synchronous speed} = \frac{\text{Cycles per second}}{\frac{p}{2}} \text{ revolutions per second,}$$

or $\dfrac{120 \times f}{p}$ revolutions per minute.

The actual speed of the motor will therefore be

$$\frac{120 \times f}{p} - \text{the slip.}$$

If the speed is altered by altering the number of poles efficient running may be obtained at either speed but the method is not flexible, that is to say, the method will give two definite speeds but does not, of itself, allow of a gradual variation of speed between them.

On the other hand, if the slip is altered, the speed variation is quite gradual and continuous but for large slips very poor efficiencies are obtained.

Regulation of speed by altering the number of poles. It has previously been pointed out that both the rotor and stator circuits must be wound for the same number of poles, and thus any alteration of the number of poles in the stator involves a corresponding

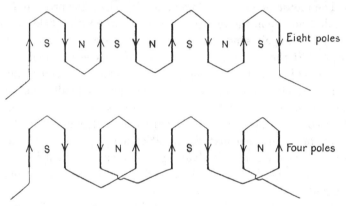

Fig. 170.

alteration in the number of poles in the rotor. If the rotor is of the squirrel cage type this is effected automatically, but if a wound rotor is used the alteration is not readily carried out and so this method is practically confined to motors having squirrel cage rotors.

The number of poles of the stator can readily be altered by providing it with two independent windings, one wound say for six poles and the other for twelve poles, in which case, if the latter winding is substituted for the former, the speed will drop to practically one-half.

Or again, the windings may be for eight and twelve poles respectively, in which case the two speeds will be approximately as 3 : 2.

In certain cases instead of having two independent windings it may be preferable to have the ends of the coils brought out to a suitable commutator, the alteration in the number of poles then being effected by a re-grouping of the coils. Thus, in the case shown in Fig. 170 (in which only one phase having one turn per coil is shown for the sake of simplicity), a comparatively simple commutating device would allow of the winding being used as a four or as an eight pole winding at will.

Alteration of speed of an induction motor by altering the slip. The slip may be increased and the speed lowered by placing resistance in the rotor circuit; this will momentarily cause the rotor current to fall with a consequent reduction in the torque.

The motor slows down thus increasing the rotor pressure (since this depends upon the relative speeds of the rotating field and the rotor) which again sends up the values of the rotor current and the torque. If the torque required at the lower speed be regarded as being equal to that at the higher speed (which will commonly be approximately the case) the motor will automatically take up a new speed, lower than the original speed, at which this result is obtained. This method may be regarded as analogous to the placing of resistance in the armature circuit of a D.C. shunt motor and it suffers from the same disadvantages, namely, poor efficiency and variation of speed as the load varies if the rotor resistance remains constant.

An alternative method of producing the required increase in slip is to lower the pressure applied to the stator, in which case the strength of the gap field decreases and the rotor current will also momentarily fall thus lowering the torque for the time being and causing the motor to slow down. As this occurs the field will still remain weak but the rotor current will rise and eventually the speed will settle down to a new, but lower value, such that the torque again rises to the value demanded by the load. If full load torque is demanded at the lower speed much more than full load current will be necessary both in the stator and rotor circuits. This is partly owing to the decrease in the field strength and partly owing to the phase relationship between field and rotor current becoming less favourable owing to the increased frequency of the rotor current (since there is no added resistance in the rotor circuit this will result in the rotor current lagging to a considerable extent behind the rotor pressure). This method is not so good as the method which involves the insertion of resistance in the rotor circuit and should only be used for motors having squirrel cage rotors and which have considerable thermal over load capacity.

At first sight it would seem that the loss of power would not be so serious in this method as when a rotor resistance is made use of, but experience has shown that there is little to choose between the two methods in this respect, possibly owing to the heavy power loss incurred on account of the large rotor and stator currents.

A combination of the pole changing method and the rotor resistance method may be employed, the former being used to obtain the large variations, and the latter the small ones. Flexibility in speed is then attained without too great a sacrifice from the efficiency point of view.

The cascade system of control for two induction motors. If two induction motors are available which can be both mechanically and electrically coupled (as, for instance, is often the case in traction work) a very interesting method is available for obtaining two economical running speeds. Suppose we have two four pole induction motors available, then, on a 50 cycle circuit, we can obtain a speed of approximately 1450 revolutions per minute (the synchronous speed would be 1500 revolutions per minute) by placing the stators in parallel on the mains and independently short circuiting each rotor. In order to obtain a lower speed the motors are re-connected (it is now essential that they be mechanically coupled) so that the stator of the first motor is connected to the line, the rotor of the first motor to the stator of the second motor, and the rotor of the second motor to the starting resistance; the machines are started in the usual way by gradually cutting out the starting resistance.

With the number of poles quoted the synchronous speed of the set would be 750 revolutions per minute and the actual speed somewhat lower. The frequency of the currents generated in the rotor of the first machine would be rather higher than 25 cycles per second and this is also, of course, the frequency of the currents supplied to the stator of the second machine. With the arrangement shown in Fig. 171 (*a*) the rotor of the first machine must either be wound to give a pressure of the same order as the line pressure for application to the stator of the second machine or a step-up transformer must be used between the two when they are in cascade.

An alternative arrangement would be to connect the rotor of the first machine to the rotor of the second machine, the starting resistance being connected to the stator of the first machine as in Fig. 171 (*b*).

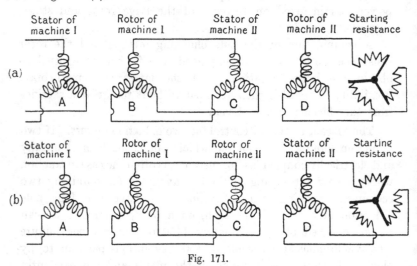

Fig. 171.

The cascade principle is also most ingeniously used in the "Cascade" motor developed by Mr L. J. Hunt and manufactured by the Sandycroft Foundry Co., to whom the author is indebted for the following information. In the ordinary application of the cascade principle there are, as will be seen from a reference to Fig. 171, four windings, and the functions of the various windings will be evident from the description already given; further, in the ordinary application of the principle, the windings are distributed over two machines, *A* and *B* being on the first machine and *C* and *D* on the second machine. In the Hunt motor these windings are all arranged on the same machine. The winding *A* is carried on the stator and produces the usual rotating field and also induces currents and pressures in the winding B which is carried by the rotor. These currents are passed on to the winding C, also carried by the rotor, which is wound for a different number of poles to that of the winding B. The combined effect of the mechanical rotation of the winding C and the alternations of the current occurring within it produces a further rotating field which is used to induce currents in the winding D (this being carried by the stator and arranged for the same number of poles as the winding C) to which the starting resistance is attached, this being cut out in the usual manner as full speed is reached. The above is a brief description of the action of the motor when working in cascade; if the higher speed is required the winding B is provided with slip rings and if the starting resistance is attached to these the motor runs up to a speed slightly below the synchronous speed corresponding to the number of poles for which the windings A and B are arranged. It is of course impossible in a work of this elementary nature to go into detail concerning the windings, which are somewhat complicated, but it will be well to point out that the windings A and D (both of which are situated on the stator) which have so far been referred to as two distinct windings are, in practice, composed to a large extent of the same conductors though the tappings for the winding A are of course taken from different points to those for the winding B.

This cascade motor has two distinct spheres of usefulness: it may be used as a two speed motor, as described above, in which case it will be necessary to provide the rotor with slip rings for use in attaining the higher speed, or it may be run as a cascade motor for a single speed only, in which case a motor is produced having the excellent starting characteristics of the slip ring motor yet having no slip rings, both the starting resistance and the mains being connected to the stationary portion of the machine. Owing to certain factors, into which it is unnecessary to enter, the cascade motor is essentially a slow speed machine, which is from some points of view advantageous, thus the motor is conveniently built with the windings A and B arranged for eight poles and the windings C and D arranged for four poles, in which case if the supply is at 50 cycles per second the speed will be about 490 revolutions per minute (the synchronous speed being that corresponding to a twelve pole motor, *i.e.* 500 revolutions per minute).

If the higher speed is attained by connecting the rotor slip rings to a starting resistance and ultimately shorting them, the result will be about 735 revolutions per minute (the synchronous speed being that corresponding to an eight pole machine, *i.e.* 750 revolutions per minute). Of course the cascade motor only gives two economical running speeds, but intermediate speeds, involving waste in regulating resistances, can be obtained by keeping resistance in the rotor slip ring circuit (for speeds between 485 and 735 revolutions per minute) or in the circuit of the stator winding D (for speeds up to 485 revolutions per minute) as the case may be.

It is stated that the torque produced by the motor is remarkably uniform at very low speeds.

Performance of Induction Motors.

Power factor. One of the chief points of interest in connection with the induction motor is the power factor and its variation with load. Of recent years considerable attention has been paid to the improvement of power factor in alternating current systems and the low power factor of the induction motor is undoubtedly one of the chief obstacles in the way of progress in the desired direction. The causes of the poor power factor are the magnetising current and magnetic leakage. The magnetising current is a wattless current and will tend to increase the angle of lag of the current taken by the motor at all loads (producing the greatest effect at light loads). Magnetic leakage, whether in the stator or rotor, causes, just as in the transformer, the windings to have a leakage reactance and the effect of this source of poor power factor will be greatest at high loads. In order that an induction motor may have as good a power factor as possible it is essential that the length of the gap be kept very small, since this will not only reduce the magnetising current but will also reduce the magnetic leakage. The general way in which power factor varies with load is shown in Fig. 172; at small loads the power factor is seen to be quite low, this may be ascribed to the large value of the wattless magnetising current compared with the load current under these conditions. As the load increases the power factor attains a maximum and then commences to fall (this point is not reached in the figure) again and this fall may be put down to the effect of magnetic leakage. The best power factor obtainable from a motor depends upon the rated speed and upon other factors to be dealt with in the design, but for normal machines it will be greater the larger the output of the machine.

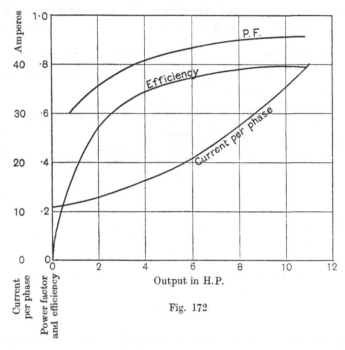

Fig. 172

Efficiency of induction motors. The shape of the curve showing the variation of efficiency with load for any one induction motor is that usual in electrical machines. Low efficiencies are obtained at low loads owing to the effect of the no load losses, consisting chiefly of the stator core loss and friction and windage loss; at some load, usually lying between half and full load, a maximum efficiency will be obtained; at still higher loads the efficiency will decrease owing to the effect of the rapidly increasing copper losses in the rotor and stator. A curve showing how the

maximum efficiency which may be expected in normal motors
varies with size is shown in Fig. 173.

Relation between slip and torque. Very interesting relationships
exist between the slip of the rotor of an induction motor and the
torque produced, and these depend to a large extent upon the
value of the resistance in the rotor circuit. If we assume that
the pressure applied to the stator has a constant value then, for
a small but constant value of the rotor resistance, increased slip
causes a practically proportional increase in the rotor current and
also in the torque so long as the total slip remains small and the

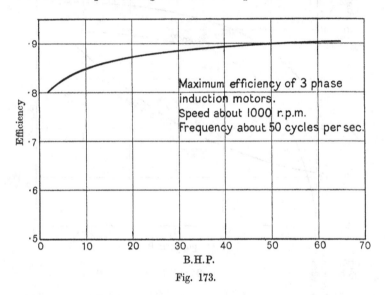

Maximum efficiency of 3 phase
induction motors.
Speed about 1000 r.p.m.
Frequency about 50 cycles per sec.

Fig. 173.

frequency of the rotor current remains low (under these circumstances the reactance of the rotor circuit has but little effect either
in limiting the rotor current or in producing difference in phase
between the rotor pressure and rotor current). As the slip
continues to increase the rotor pressure continues to increase in
proportion, but the rotor current will increase less than proportionally owing to the increasing value of the rotor reactance, the
angle of phase difference between the rotor current and pressure
will also continue to increase and the torque will therefore increase
at a less proportional rate than the current. For still greater

values of the rotor slip the current will continue to increase but at a smaller and smaller rate, while, as regards torque, the increasing difference of phase between the rotor current and pressure will ultimately compensate for the increase in the rotor current and the torque therefore attains a maximum value and will fall for greater values of the slip. These points are well brought out in Fig. 174, which also shows the effect of added rotor resistance.

With added rotor resistance the same general relationships exist between the quantities under consideration but the slip attains

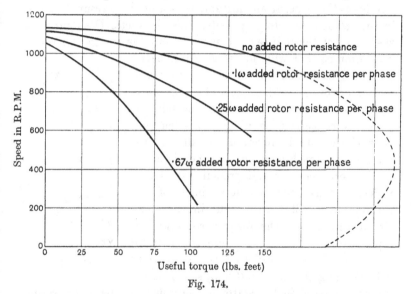

Fig. 174.

a greater value before the maximum value of the torque is reached owing to the effect of the added resistance in limiting the current for a given slip and, at the same time, maintaining a more favourable phase relation between the rotor current and pressure (*i.e.* between the rotor current and the field in the gap of the motor).

Action of the induction motor at speeds above synchronism. The vector diagram of an induction motor which is running with some slight slip (*i.e.* just below synchronous speed) is shown in Fig. 175, where E is the pressure applied to one phase of the stator and I_m is the magnetising current. The absolute frequency

of the rotor currents is of course low, but, if the frequency of the
rotor currents relative to the stator conductors is considered, we
must not only take into account the absolute frequency of the
rotor currents but also the frequency with which the bands of
rotor current move relative to the stator conductors owing to
the mechanical motion of the rotor. If a little thought is expended
on this matter it will be realised that the frequency of the bands
of rotor current and pressure sweeping by the stationary stator
conductors is the same as the frequency of the supply current,
and, further, the direction of the induced pressure in the rotor conductors is the same, at any instant, as that of the pressure induced
in the stator conductors opposite, thus the vector OB represents
the phase of the induced pressure in both the stator and rotor.

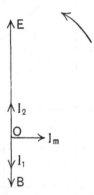

Fig. 175.

The rotor current will, for small slips, be represented by I_1 and this will cause, by transformer action, an additional stator current I_2
which, so far as the pressure induced in the
stator is concerned, is a motoring current.
Now if the slip is reversed, by driving the
rotor above synchronous speed, the directions
of the rotor current and pressure will be reversed because the magnetic lines of the
rotating field will then be cut in the opposite direction, and the rotor current will be
represented by I_2. The additional stator
current which is induced by this rotor current
will be along the line of B and, so far as the
pressure induced in the stator is concerned, will be a generating
current, the machine therefore acts as a generator and the faster
we drive it the more powerfully will it generate. It should be
noted that the frequency of the current supplied is settled by
the frequency of the magnetising current, which is still supplied
into the stator, and is independent of the speed at which the
rotor is driven—a valuable property of this device which is
sometimes on this account termed an asynchronous generator. It
is necessary that an ordinary synchronous generator be operated
in conjunction with asynchronous machines in order that it may
supply the necessary magnetising current and set the frequency.

The property of the induction motor in acting as a generator when driven above synchronous speed makes it possible to use the induction motor for regenerative braking and this has been taken advantage of in making use of such motors for electric traction on railways.

The Single Phase Induction Motor.

If a motor of the same general type as those described above is provided with a single phase stator winding, instead of a two or three phase winding, then (apart from the influence of rotor currents) we shall have an alternating magnetic field produced in the gap and at first sight it is not at all clear how motion of the

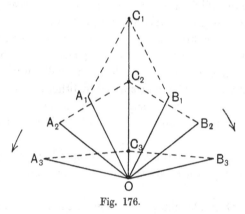

Fig. 176.

rotor may in this case be obtained; in order to arrive at a general understanding of the properties of such a motor a simple artifice is best made use of. A study of Fig. 176 will show that the resultant of two rotating magnetic fields of equal magnitudes, rotating with equal frequencies in opposite directions, is a simple alternating magnetic field having a frequency equal to that of either of its components and a maximum value equal to the sum of the magnitudes of its components. In the figure the lines OA_1, OA_2, etc. represent successive positions of one of the rotating fields, OB_1, OB_2, etc. representing the simultaneous positions of the other rotating field, the corresponding values of the resultant being denoted by the lines OC_1, OC_2, etc.

If we consider the reverse of the above theorem we can represent

a single alternating magnetic field by two equal but oppositely
rotating components, a type of magnetic field with whose properties we are well acquainted. Each of the imaginary oppositely
rotating fields which we regard as substituted for the alternating
field may be looked upon as exerting its own torque upon the
rotor, and, when the rotor is at rest, the torques will be equal and
in opposite directions, the resultant torque on the rotor being
zero. Now if the rotor is turned in one direction by some
external means the torque due to the forward rotating field will
increase, and that due to the backward rotating component will
decrease, provided that the added resistance, if any, in the rotor
circuit is not too high, as indicated in Fig. 174, and if the initial
speed is made sufficiently high the difference between the torques
will be large enough to cause the motor to run up to a speed but
slightly below synchronism in the direction in which it has been
started. We see then that the single phase induction motor will
not be self-starting in its simple form but will continue to run in
whichever direction it may be started by other means. The above
method of treatment may perhaps seem somewhat artificial but
it has the merits of simplicity and of explaining the observed
results.

When the motor is running at full speed a rotating field can
be shown to be produced by the interaction of the stator and rotor
currents.

Starting of single phase induction motors. In order to start
induction motors of the single phase type they are frequently
provided with a special starting winding which is arranged on
the stator in a position electrically at right angles to the main
winding, i.e. in the position which would be occupied by the second
winding of a two phase motor. If through this starting winding
a current is passed, which differs considerably in phase from the
current taken by the main winding, a crude form of rotating field
will be produced which will provide sufficient torque for starting
under no load or even under a small load.

The difference in phase between the currents in the main and
starting phases is obtained by means of what is termed a phase
splitting device, several modifications of which will occur to the
reader. One method is to place the main winding directly on the

supply pressure and the starting winding on to the supply pressure
in series with a resistance. The current in the former will lag considerably behind the pressure, while the current in the latter will
lag to a much less extent, and thus the required difference in phase
is obtained. The rotor of the motor may be of the squirrel cage
type or it may be wound for three phases in order that a starting
resistance may be inserted to limit the starting current taken by
the motor, and, at the same time, improve the starting torque.
A diagram of the connections of such an arrangement is shown

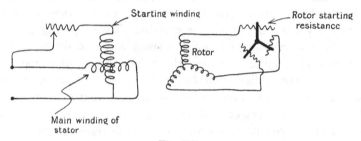

Fig. 177.

in Fig. 177, and it is clear that the starting winding on the stator
may be open circuited when the motor has attained full speed.

Another interesting and useful method of starting a single phase
induction motor is to provide the rotor with a winding of the direct
current type and with a commutator. Short circuited brushes are
placed on the commutator and the machine is started up as a
single phase repulsion motor. When full speed has been attained
the commutator segments are short circuited, either by hand or
by an automatic centrifugal device, and the machine continues to
run as a single phase induction motor having a short circuited rotor.

CHAPTER X

CONVERTING PLANT

The term converting plant is commonly applied to rotating machines or other appliances arranged to convert power from alternating to direct current or *vice versâ*, though they are most frequently used for the former purpose. Many lighting schemes in this country started in quite a small way and direct current satisfied all requirements from the points of view of generation, utilisation and transmission, but, with the rapid growth of such schemes, difficulties in the sole use of direct current soon appeared. In most cases direct current supply is still the most convenient so far as the consumer is concerned* but in regard to generation and transmission there are, at present, considerable advantages to be obtained by the use of alternating current.

Alternators are successfully made in much larger sizes (for direct coupling to steam turbines) than are direct current generators, and the use of these large units leads to economy in floor space and increased efficiency. Again, with the areas supplied by many lighting schemes growing so large, it has become essential to use high pressure in feeding to the outskirts in order to secure economical transmission, and these high pressures are, at the present time, more readily obtained with alternating than with direct current†.

In this country a very common compromise between alternating and direct current for lighting schemes consists of making use of

* This remark is intended to apply to systems which are essentially for the supply of energy for lighting purposes and to which a small motor load is attached, it does not apply to the supply of energy to textile mills or to supply of energy in bulk in which cases three phase alternating current is usually more convenient from all points of view.

† It is worthy of mention that at the present time the Thury system is being developed for the purpose of transmitting power by high tension direct current but it is too early to state whether this system is likely to make any considerable progress.

a three phase central generating station and a three phase transmission to sub-stations situated in convenient distributing centres. In these sub-stations converting plant (often of very considerable amount) is installed for the conversion of the power to direct current for lighting and traction purposes. Where alternating current is supplied to the individual consumer it may be necessary for him to convert to direct current for special purposes as, for instance, charging accumulators, but in such a case a type of device other than a rotating machine is perhaps preferable. The importance of installing converting plant with a high efficiency cannot be overrated and it will often be advantageous to incur extra initial outlay in order to secure a slight gain in efficiency*.

Of course this remark also applies to other plant and unfortunately undue importance is too often attached to economy in first cost, whereas the lowest overall financial results are often attained by using apparatus which in the first place costs rather more but which effects a saving in running cost. Thus, to take a simple example, let us consider two 1000 K.W. sets running for six hours per day at full load and let the difference in efficiency be 1 %, that of A being 93 % and that of B being 94 %. Taking the cost of energy wasted as ·5 penny per unit the annual cost of the energy wasted by A will be $\dfrac{6 \times 365 \times \cdot 5 \times 75 \cdot 2}{240}$ or £343, and that by B will be $\dfrac{6 \times 365 \times \cdot 5 \times 63 \cdot 8}{240}$ or £291.

The difference is £52 per annum which, capitalised at 5 %, represents £1040, and it will be well worth while to spend a part of this initially in order to obtain the higher efficiency plant.

The chief methods available for converting from alternating to direct current are by the use of:

(1) A synchronous motor generator,

(2) An asynchronous motor generator,

(3) A rotary converter (usually called a rotary),

(4) A motor converter,

(5) An electric valve.

---

* If the plant is likely to work on a variable load it is of course the all day efficiency which should be considered.

The asynchronous motor generator. This consists of an induction motor directly coupled to a direct current generator mounted on the same bed plate.   There is little that need be said in regard to this type of machine since the induction motor has already been dealt with and the reader is supposed to be familiar with the operation of direct current generators; the only points that need be considered are those concerning its operating merits.

The set needs a comparatively large floor space, since it comprises two independent machines, but has the merit of allowing the direct application of high alternating pressure to the stator of the induction motor, thus dispensing with the need of transformers and with the accompanying additional switch gear.   It is self starting from the alternating current side but, unless auxiliary appliances are fitted, the power factor will always be below unity (the current lagging) and this will tend to cause poor regulation on the part of the alternators from which the set is run.   On the direct current side a wide range of pressure is readily attainable by the ordinary method of regulation of the generator field.

The synchronous motor generator. This again is composed of two distinct machines the driving motor this time being of the synchronous type and the other machine an ordinary direct current generator as before.

Before dealing with the operating properties of this type of plant it will be necessary to consider the action of the synchronous motor and the following consideration applies not only to motors forming part of converting sets but also to synchronous motors used for other purposes.

Imagine an electro-magnet consisting of a laminated iron core surrounded by a winding through which an alternating current can be passed, and immediately in front of this a pivoted magnet which may either be permanent or of the electro type (see Fig. 178).   If the movable magnet is at rest in some such position as is shown in the figure, and the stationary electro-magnet excited with an alternating current, the mechanical forces acting on the movable magnet will be of an alternating nature as the polarity of the fixed magnet reverses, *i.e.* during one half cycle N will be attracted by A, and during the next half cycle it will be repelled with an equal force, and the nett result will be at the most a slight

oscillation of the movable magnet.   Now suppose that the movable
magnet is caused to revolve at the same number of revolutions
per second as the frequency of the alternating current supply
(this remark only applies to the two pole case shown in the diagram),
and that current is switched on to the alternating current magnet
in such a phase that the polarity of the stationary magnet becomes
approximately zero when stationary and moving poles pass each
other, and also so that when a moving N pole is approaching one
of the fixed poles the latter is of south polarity.

A study of Fig. 179 will show that the mechanical forces exerted
on the moving magnet are now always in the direction of rotation,
though in amount they vary between zero and a maximum, and

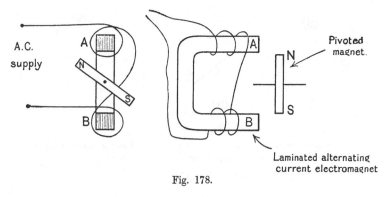

Fig. 178.

we see at once that unless the retarding forces, due to friction or
other causes, are too great the movable magnet will continue to
revolve and, further, it will revolve exactly in step with the
alternations in polarity of the fixed magnet.   If it lags behind
the position corresponding to synchronous speed by more than
a small fraction of a revolution the resultant force tending to
drive the movable magnet will become zero and it will rapidly
come to rest.   On account of the fact that the movable member
must rotate synchronously this type of motor has received the
name of a synchronous motor and it is clear that in its simple form
it will not be self starting.

If we consider an ordinary single phase machine, constructed
in exactly the same manner as a single phase alternator, in which
the stator is supplied from alternating current mains with

current of suitable frequency and pressure, and the rotating field magnet is excited with direct current, and if this is run up to the correct speed by some external means and synchronised with the alternating current mains, we have a device precisely similar (except of course in regard to details of construction) to the simple motor dealt with in the above consideration; we thus realise that a synchronous motor is constructionally an ordinary alternator which is driven electrically and which generates mechanical power, instead of being driven mechanically and generating electrical power as would be the case if it was used in the usual way.

If the machine is for two or three phases we have practically the same conditions but in this case the armature currents will

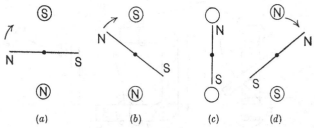

Diagram showing forces developed between fixed and moving magnets
for successive positions of the latter.

Fig. 179.

produce a rotating rather than an alternating field and the field system, excited with direct current, will follow this field synchronously. Whether the machine is for single or multiphase we shall have a synchronous motor which is not self starting unless special arrangements are made as described later.

The torque obtained in a multiphase machine will be practically uniform whereas that obtained from a single phase machine will be of a pulsating nature, in fact it is very likely that a reverse torque will occur during certain parts of each cycle.

Having examined the fundamental idea underlying the operation of the synchronous motor, let us next go into greater detail with the aid of a vector diagram. Imagine that a synchronous motor (either single or multiphase, though of course the vector diagram will deal with one phase only) is run up to speed by some external means, its field excited so that the pressure generated in the armature winding is equal to the pressure of the line, and the armature synchronised with the supply mains. In Fig. 180 (a) the vector OA represents the line pressure and OB the pressure generated in the armature of the motor, and, since these are equal in magnitude and opposite in phase, no current will flow and if the external driving force is removed the motor will at once commence to slow down. When it has dropped a small fraction of a revolution behind the position corresponding to synchronous speed, the vector OB takes up the position shown in Fig. 180 (b)

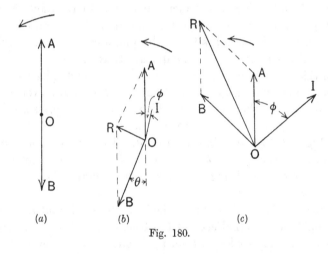

Fig. 180.

and we see that the line and motor pressures no longer neutralise each other but have a resultant, OR, which is free to send a current round the armature circuit, the path being completed by the mains, bus bars, and the generator armatures attached thereto. The impedance of this circuit will be chiefly due to the resistance and reactance of the motor armature and, since the latter will in most cases predominate numerically, the resulting current, shown in the figure by OI, will lag considerably behind the resultant pressure. It will be seen that as regards the mains pressure this is a generating current while as regards the machine pressure it is a motoring current, the machine thus takes power from the mains and will continue to run synchronously but with its pressure a little behind true phase opposition to the mains pressure. If the machine drops back a little more relative to the mains pressure, the resultant pressure, the current, and the power taken from the mains will all increase. For any load on the motor, within limits depending upon the capacity of the machine, the vector representing the machine pressure will drop back relative to the vector representing the mains pressure until the input of the motor is just sufficient to keep it running against the load.

It should be noted, however, that if the motor pressure drops back to too great an extent relative to the mains pressure, as in Fig. 180(c), though the armature current continues to rise, the input power falls off owing to the increasing value of the angle ϕ (the input power is $IE \cos \phi$, E being the applied pressure) and it follows that if the motor is loaded beyond a certain critical value it will fall out of step and come quickly to rest.

In this connection Fig. 181, in which, for a single phase motor, we have input current, input and output watts, and power factor plotted against the angle by which the motor pressure lags behind true phase opposition to the mains pressure (*i.e.* the angle θ in Fig. 180(b)), will be of interest. The applied and motor pressures are each taken as 170 volts at 50 cycles per second, the armature resistance being ·6 ohm, and the reactance 2·34 apparent ohms.

The curve labelled output in the figure is not the output which is available for external use but has been obtained by subtracting the armature copper losses from the input. The ordinate of this curve may be regarded as proportional to the total torque of the machine, some of which will be used internally on account of friction and core losses.

It will be seen that the maximum total torque is obtained when the angle of the motor pressure behind true phase opposition is about 75° and if the load is such as to demand a greater torque than is then given by the motor it will fall out of step and come to rest. The portions of the curves shown by continuous lines represent conditions which may be attained in the normal operation of the machine, but the dotted portions represent unstable conditions which are only passed through as the machine is falling out of step.

Effect of varying the excitation of a synchronous motor. For convenience and in order to simplify the matter let us consider that the phase difference between the line pressure and the pressure generated in the motor is constant, then, if the exciting current of the field magnet is altered, the magnitude of the pressure generated in the motor will also be altered*.

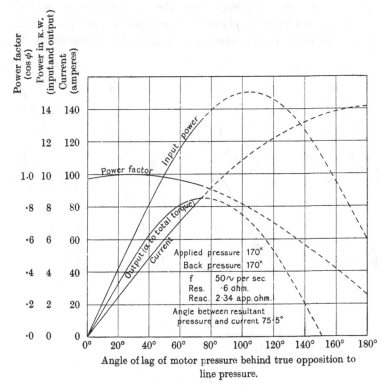

Fig. 181.

In Fig. 182 vector diagrams are given showing the phase relations of the various quantities concerned for different values

---

* As a matter of fact in practice the action is more complicated than would appear from the above statement. The field flux is really due to the resultant of the field ampere turns and of certain ampere turns due to the wattless component of the armature current, and if, for example, the field ampere turns are increased, the effect is partially neutralised by a counter effect due to the wattless component of the armature current.

of the back pressure generated by the motor (*i.e.* for different values of the field current). At (*a*) we have the motor excited to a pressure equal to the mains pressure and the current is seen to lag slightly behind the applied pressure; at (*b*) the excitation of the motor has been lowered and the current lags still more behind the applied pressure; at (*c*) the motor is excited so that the back pressure is greater than the applied pressure and the current is seen to be in phase with the applied pressure; finally, at (*d*), the motor is considerably over excited with the result that

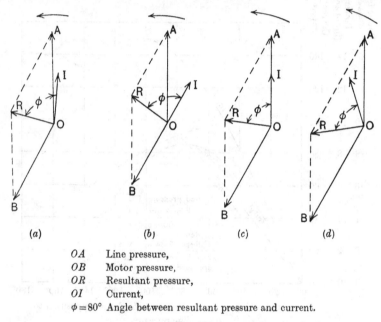

OA	Line pressure,
OB	Motor pressure,
OR	Resultant pressure,
OI	Current,
$\phi = 80°$	Angle between resultant pressure and current.

Fig. 182.

the current taken actually leads the applied pressure. It will be seen from the above considerations that the power factor of a synchronous motor is adjustable, between wide limits, to any desired value and by very simple means, namely, varying the direct current excitation of the field. This is a most valuable property and advantage may be taken of it in order to adjust the power factor of the motor to unity at any load, an operation that is impossible with the ordinary induction motor. Further, if desired,

the synchronous motor may be caused to take a leading current
and in this way compensate for lagging current taken by other
devices, as induction motors, run from the same mains. Curves
which were obtained from a 12 H.P. three phase machine, showing
the variation of power factor, watts wasted in the machine, and
armature current, caused by varying the field current, are shown
in Fig. 183.

Hunting of synchronous motors. Consider a mechanical coupling
comprising two discs which are connected by springs as shown in

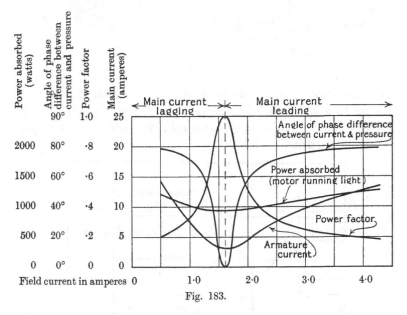

Fig. 183.

Fig. 184, the disc *A* having some load attached to it and being
driven by the other. It will be obvious that the springs must
extend before mechanical power can be transmitted through the
coupling, and the greater the power to be transmitted to the disc
A from the disc *B* the greater must be the extension of the springs.
If the load on the driven disc is suddenly increased, a momentary
slowing down of this disc will result, and at first sight it might be
thought that the driven disc would run rather slower until the
springs had increased their tension sufficiently to provide the
additional torque, after which the two discs would again rotate at

equal speeds. As a matter of fact, however, when the spring has
extended to the required amount the driven disc will be going too
slow, and during the process of speeding this disc up to the speed
of the driving disc the spring will become extended to an amount
that is greater than that necessary to cope with the steady load

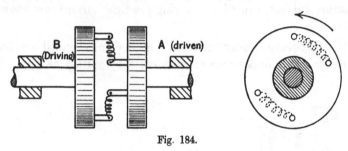

Fig. 184.

that has been put on. The result is that the driven disc will be
run up to a speed slightly greater than that of the driving disc
and the spring will become less extended than is necessary to supply
the torque to drive the load.

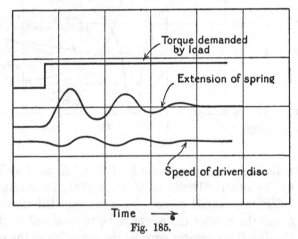

Fig. 185.

The driven disc then commences to slow down again and this
oscillation of the driven disc relative to the driving disc will
continue, with a gradually decreasing amplitude, until the energy
of the oscillations is completely damped out by friction. These
oscillations are shown graphically in Fig. 185.

The period of the oscillation will depend upon the strength of the springs and upon the inertia of the driven portion. These phenomena are said to be due to "hunting" or phase swinging of the driven disc and in the ordinary course of events will die away fairly quickly, but, if the disturbing cause is periodic and has a period approximately equal to the natural period of swing of the second disc relatively to the first, a species of resonance might ensue resulting in oscillations of very large amplitude and possible breakage of the springs.

In a synchronous motor (and also in other plant as rotaries and motor converters) a somewhat similar state of affairs may exist. The coupling between the rotor and stator is not of course in this case a mechanical spring, but is what may be described as an electro-magnetic spring, which is distorted to a greater or less extent according to the load on the motor*.

When load is put on to the rotor the angle of phase difference between the applied and generated pressures in the armature circuit of the motor has to adjust itself to the new condition of load (just as .the spring did in the mechanical case) and instead of taking the right value immediately it overshoots the mark. In other words the rotor may, as a result of an alteration of load (or indeed as a result of other causes as alteration of supply frequency, drop of line pressure, etc.), have superposed on its uniform angular velocity an oscillatory motion. This phase swinging causes a periodic alteration of the phase angle between the applied and generated pressures in the armature circuit of the motor and this gives rise to fluctuations of the resultant pressure and current taken by the motor, and the swinging of the ammeter needle is the most convenient way in which the presence of hunting may be detected. In general this oscillatory motion will soon die away as the result of eddy currents set up in pole pieces, but, if the disturbing cause has a period approaching the natural period of oscillation of the rotor, the magnitude of the oscillations may gradually increase until at last the machine falls out of step and comes to rest (corresponding to the breaking of the spring in the mechanical case).

* The spring effect is of course due to the mechanical forces developed between the armature currents and the poles of the field system.

In order to encourage the damping out of phase swinging by
means of eddy currents, the pole pieces of synchronous motors
often have copper bars placed through them which are short
circuited at each end by other bars, thus forming a grid of high
conductivity in which the necessary eddy currents are formed to
a maximum extent. If a flywheel is attached to a machine
which gives trouble due to hunting a cure can sometimes be
effected by removing or altering the wheel. Any alteration of
the inertia of the wheel results in an alteration of the natural
period of the rotor and thus may prevent synchronisation with
the disturbing cause which is possibly periodic.

Starting of synchronous motors. In the past, starting has usually
been effected by means of an auxiliary motor, the main motor
being run up to speed, its field excited, and then synchronised on
to the mains in the usual way. One method of carrying this
into effect has been to make use of the exciter (for supplying the
field current) in conjunction with a small battery of accumulators;
the battery is charged from the exciter while the main motor is
running and when the main motor needs re-starting the exciter
is for the time being run as a motor, power being obtained for this
purpose from the battery. Another method is to directly couple
a small induction motor on to the shaft of the synchronous motor;
if the induction motor is wound for one pair of poles less than the
main motor there is no difficulty in running the synchronous
motor up to a speed high enough for synchronising, a fine regulation
being obtained if necessary by a resistance placed in the rotor
circuit of the induction motor. It is clear that expense would in
general prohibit the use of anything but a small auxiliary motor
and it will therefore be necessary to remove the load from the
main motor before starting up. A third method of starting up
three phase synchronous motors which has attracted considerable
attention consists of switching on the line pressure to the stator,
the field being unexcited. A rotating field will be set up by the
armature currents, and eddy currents will be induced in the pole
pieces and damping grids (if any) and the machine starts up as
an induction motor; when a speed near to synchronism (but of
course slightly below) is reached the field current is switched on
and the motor pulls into step.

This method has been successfully developed by the Lancashire Dynamo Co. (to whom the writer is indebted for the following information and for Fig. 186) in their Lancashire self starting synchronous motor. The stator of this motor is of ordinary construction and the rotor is of the smooth cylindrical core type, such as is often used for the rotors of turbo alternators, with a partially distributed field winding. The starting arrangements are shown in Fig. 186, from which it will be seen that when the supply is switched on to the stator and the starting handle moved from the off position to the first contact, the rotor field winding is disconnected from the exciter and connected to a starting resistance and the machine runs up to speed (as resistance is gradually

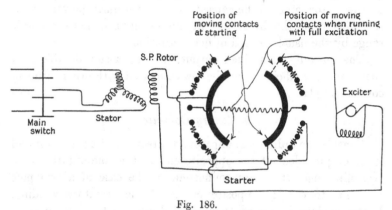

Fig. 186.

cut out of the rotor circuit) as an induction motor having a single phase rotor.

When the starting resistance in the rotor circuit has been completely cut out and the rotor has attained a speed slightly below synchronism, a further movement of the starting handle temporarily reintroduces a buffer resistance and then switches the exciter into the circuit in series with a resistance. The exciter has already developed a pressure owing to the rotation, and thus direct current is switched on to the field circuit and the main machine locks into step and continues to run synchronously. Further motion of the starting handle increases the strength of the field current and may be used for the purpose of adjusting the power factor.

It is important to notice that the connections shown ensure
that the starting operations are carried out in the correct order.
It is not to be expected that a starting torque equal to full load
torque should be readily attained but in practice it has been found
possible to obtain sufficient starting torque for direct coupled fans,
pumps, etc. It should be clearly understood that it is not only
possible to construct such machines to run at unity power factor
but also to make them, if desired, take a leading current which
can be used to compensate for lagging current taken by other
machines. We are now in a position to consider the synchronous
motor generator set as used for converting plant; it can be made
self starting, can be arranged to take high tension current directly
on the stator, and can be caused to run with unity power factor,
the pressure on the direct current side being variable over a wide
range by the usual method of field variation.

The set will, for a given output, take up a considerable floor
space, and the efficiency will be low compared with other forms of
converting plant.

The Rotary Converter.

Consider an ordinary shunt wound direct current generator and
let it be provided, at the opposite end to the commutator, with
two slip rings which are connected, in the case of a two pole
machine, to exactly opposite points of the armature winding.
If the machine is run up to speed and allowed to excite itself it
is clear that we shall be able to obtain direct current from the
commutator and, at the same time if necessary, single phase
alternating current from the slip rings. Such a machine is termed
a double current generator. It is desirable to point out that so
far as the pressure generated in the individual conductor is concerned it is quite immaterial whether the machine is supplying
a direct or alternating pressure, the pressure in the individual
conductor is an alternating one and will have the same wave form
in either case, this depending upon the field distribution. If the
armature conductors are permanently connected to the exterior
circuit in any particular way (as when slip rings are employed)
we obtain in the latter an alternating pressure, whereas if a
commutator is employed between the conductors and the external

circuit we have the alternating pressures produced in the armature conductors rectified and so obtain a uni-directional pressure in the external circuit. It should be noted that the wave form of the *current* in the conductor will not be the same in the two cases since this will depend not only upon the pressure generated in the conductor in question but also upon the pressures generated in conductors which are, for the time being, connected in series with the conductor under consideration. Fig. 187, which is drawn for a conductor situated midway between the connections to the rings and for a non-inductive external circuit, shows the relative shapes of the wave form of the current in the conductor when direct current is being taken from a commutator and when

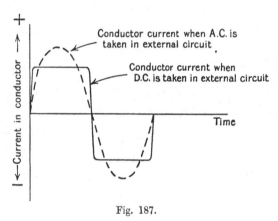

Fig. 187.

alternating current is being taken from slip rings respectively. If the external circuit is not non-inductive the only difference will be (assuming that the same current, as read on a hot wire ammeter, is taken in each case) that the wave for the slip ring case will be moved along the axis of time one way or the other.

If the machine is run as a motor from the direct current side it will clearly be possible to obtain alternating current from the slip rings, and thus it is possible to convert from direct to alternating current by using a single rotating machine. Again, the same machine may be run as a motor (of the synchronous type) from the alternating current side, if it is first started up and synchronised in the usual manner, and when running it will

be possible to obtain direct current from the commutator*; a
machine running in this way (*i.e.* motoring on the A.C. side and
generating on the D.C. side) is known as a rotary converter or more
shortly as a rotary. When motoring from the D.C. side and
generating on the A.C. side it is known as an inverted rotary.
If it is desired to make use of three phase current, three slip rings
will be used, connected, in the two pole case, to three equidistant
points of the armature winding. A completely distributed winding,
such as we have had in mind, is not ideal from the alternating
current point of view, but of course considerations arising
from the direct current side render the use of such a winding
necessary.

If the machine is used as a double current generator the

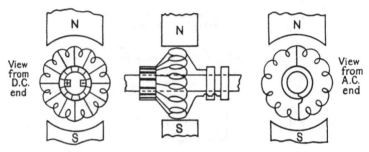

Diagrammatic representation of a rotary converter.

Fig. 188.

direction of the currents in the armature conductors due to the
D.C. output will be *generally* the same as those due to the A.C.
output; but if one side of the machine is motoring and the other
is generating the directions of currents due to the two sides will
be generally opposite†.

---

* The direct current for the field system can of course be obtained from the
commutator instead of from a separate exciter as would be the case with an ordinary
synchronous motor.

† This statement should not be taken too literally since the phase of the current
in the armature conductors due to the A.C. side will depend very largely upon the
power factor and also upon the position of the conductor referred to relative to the
slip ring connections. The magnitude of the current on the A.C. side relative to
that due to the D.C. side will also depend upon the power factor on the A.C. side,
but when the machine is used as a converter there will, in general, be considerable

Relationship between the direct and alternating pressures. To investigate this matter let us consider the induced pressures on the two sides in order to simplify the problem, the results obtained will be generally applicable to the corresponding terminal pressures though the ratios between the latter quantities will be slightly affected by the internal pressure drops due to armature resistance and reactance.

In considering the relative values of the induced pressures on the alternating and direct current sides it is clear that the pressure induced between any two points of the winding will be the vectoral sum of the individual pressures in the conductors lying between the two points in question. Since the winding is a distributed one there will be a small phase difference between the pressures occurring in adjacent conductors, and if a complete vector diagram is drawn showing the addition of the pressures occurring in all the conductors of the armature the resultant will be zero and the shape of the vector diagram practically a circle since the individual pressures will be very numerous and small. In Fig. 189 the steady D.C. pressure will be represented by the line AB (in a two pole case where the D.C. pressure will be the sum of the pressures produced at any instant in one half of the total conductors). Again, if B and C represent

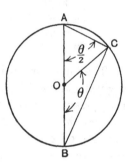

Fig. 189.

the points of connection of two slip rings on the A.C. side, the maximum value of the alternating pressure produced between them will be represented by the line BC and the R.M.S. value of this pressure can be found by dividing the maximum value by $\sqrt{2}$ if the pressure is assumed to be sinusoidal. It is now quite a simple matter to calculate the relative values of the induced pressures on the A.C. and D.C. sides if the angle between the slip ring connections is known and this will be settled by the number of phases for which the machine is intended on the A.C. side.

neutralisation of the two sets of current in the armature conductors leading to diminished heating and armature reaction.

R.M.S. value of pressure between rings

$$= \frac{BC}{\sqrt{2}} = \frac{AB \sin\frac{\theta}{2}}{\sqrt{2}} = \frac{\text{D.C. pressure} \times \sin\frac{\theta}{2}}{\sqrt{2}}.$$

All through this investigation we have had in mind the case where a two pole magnetic field is used but the final results are equally applicable to machines having any number of poles.

Table of values of alternating pressures in converters, direct steady pressure taken as 100.

No. of phases	Electrical angle between adjacent slip ring connections	R.M.S. alternating pressure between adjacent rings
Single phase	180°	70·7
Three phases	120°	61·2
Six phases	60°	35·3

The important point to note is that the A.C. pressure is, for a given number of phases, fixed within very small limits by the D.C. pressure or *vice versâ*, and since D.C. pressures for lighting and traction are commonly not more than 500 or 600 volts, the pressure on the A.C. side is also relatively low. It is impossible therefore to connect the slip rings directly on to high tension lines, and step down transformers must always be used between the high tension or extra high tension lines and the A.C. side of the rotary converter.

The rotary, like other electrical machines, is self regulating as regards the output and input, that is to say the input always adjusts itself to the output, due allowance being made for the losses occurring in the machine.

It is possible, therefore, to calculate approximately, for a rotary of a known number of phases and working under known conditions, the relative values of the currents on the D.C. and A.C. sides, the exact ratio of course depending largely on the power factor on the A.C. side.

As a simple example consider the case of a single phase rotary converter; neglect the losses occurring in the machine, and assume

that the power factor on the A.C. side is unity; the volt-amperes on
the D.C. side will then be equal to the volt-amperes on the A.C. side,
and since the alternating pressure is ·707 of the direct pressure it
follows that the alternating current will be 1·414 times the current
on the D.C. side. It will also be clear that the maximum value of
the current on the A.C. side will be twice that of the current on
the D.C. side. In a similar way the theoretical current ratios in the
case of multiphase converters can be determined since, if there
are n phases, $\frac{1}{n}$-th of the total power will be dealt with by each
phase on the A.C. side; by dividing this power per phase by the
phase pressure, the phase current is obtained, and, finally, the line
current can be found by combining two phase currents at the
correct angle of phase difference (in the three phase case this will
be 60°, see page 86).

Wave form of current in the armature conductors. Since the
conductors are traversed by both alternating and direct currents
the wave form will clearly be somewhat complicated. In Fig. 190
wave forms are shown for three different conductors in a single
phase rotary, the power factor being taken as unity, in which case
the maximum value of the alternating current will be twice that
of the direct current. Each of the curves starts from the instant
when the slip ring connection is passing the brush, and whether
the conductor considered is situated at (a), (b) or (c) the phase of
the alternating component is the same and thus this component is
in each case plotted in a similar position. The point at which
the direct current reverses will depend upon the position of the
conductor under consideration relative to the slip ring and will of
course be later for (b) than for (c), and later still for (a). The resultant
curve is in each case obtained by adding the two components at
any instant algebraically and is shown by the thick line.

It will be seen from the diagram that the R.M.S. value of the
conductor current differs considerably from coil to coil and the
heating of the different coils will therefore be far from uniform.
Similar wave forms would be obtained if a three phase case was
under consideration but the heating, though still dependent on
the position of the coil relative to the slip rings, would not vary

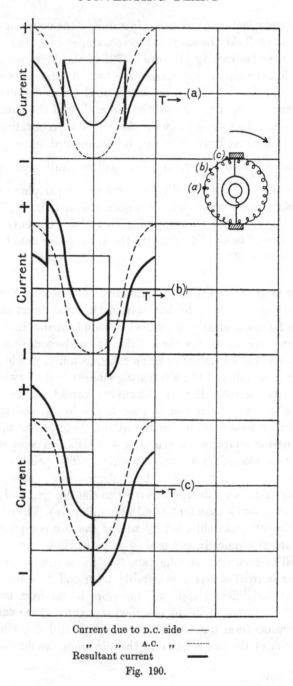

Current due to D.C. side ——
    „   „   A.C. „ --------
Resultant current ——
Fig. 190.

between such wide limits as in the single phase case; the average heating would be much less than in the single phase case, and also less than would occur if the same machine was used to give the same output as a simple generator either for direct or alternating current. The output of a certain carcase, used as a rotary converter, is greater, from the heating point of view, than when used as a simple generator except when the rotary is used as a single phase machine. In practice single phase converters are extremely rarely met with but two phase (four wire), three phase and six phase machines are very commonly used.

Starting of rotary converters. This subject has of late attracted considerable attention and the ideal method of starting is one which combines as far as possible the following points:

(1) Low starting current from mains and absence of current surges.

(2) Simplicity in arrangement and, more particularly, in operation.

(3) Quickness in getting the machine on to load.

(4) Certainty in producing correct polarity on D.C. side.

Starting from the D.C. side. To effect this a motor starter is provided on the direct current side and the machine is started up as a direct current motor. There are two objections to this method, the first being that a supply of direct current is necessary and even if this is available normally it might possibly be absent in the event of a bad shut down in the sub-station and thus, when it is perhaps most necessary and urgent to start up rotaries quickly, the means for so doing might not be available; the second objection is that synchronising is necessary on the A.C. side, but the correct D.C. polarity is of course assured.

Starting from the A.C. side.

(a) *With the help of a separate alternating current starting motor.* This is carried out by directly coupling to the rotary a separate starting motor which is usually of the induction type, though a commutator motor may replace it if desired. If of the induction type it is usually arranged for two poles less than the rotary in order that there may be no difficulty in reaching the synchronous speed. This method involves synchronising on the A.C. side, but the correct polarity is assured on the D.C. side owing to the residual magnetism of the field magnets. It is a method largely used in practice.

(b) *Starting up the rotary as an induction motor.* If a fraction of the ordinary pressure is supplied to the slip rings of a rotary, say by means of tappings on the low tension side of the transformer, a rotating magnetic field will be produced by the armature, and if the pole pieces are provided with short circuited rings or grids (which will also serve to prevent hunting) eddy currents will be induced in them and the machine will run up to a speed just below synchronism as an induction motor. If now the switch in the direct current exciting circuit (previously kept open) be closed, the machine will pull into step and continue to run synchronously; the full alternating pressure may then be applied, the direct pressure adjusted, and the machine put on load. The machine is, in fact, a self synchronising converter, and this method of starting, though simple in principle and operation, has several disadvantages. In the first place the starting current taken by the machine is considerably in excess of the full load current, though it is only fair to point out that the current taken from the line may be less than full load current owing to the step down ratio of the transformer being greater than in ordinary running; again, a considerable current surge may take place when the machine pulls into step; further, the polarity produced on the D.C. side cannot be depended on since the residual magnetism is wiped out by the armature reaction produced by the considerable starting currents used; just before synchronism is attained each pole will alternately be magnetised of north and south polarity as the poles of the armature slowly slip by the poles of the field and the final polarity attained depends upon the instant at which the field switch is closed*; finally, sparking on the commutator may be very severe during the process of starting up owing to the large currents produced in the coils shorted by the brushes.

* If the polarity comes up the wrong way round it may usually be readily rectified by opening the field switch for a short period thus causing the machine to slip a pole.

(c) *Self synchronising converter using separate starting motor.*
The British Westinghouse and other companies have recently
produced a self synchronising converter in which the disadvantages
mentioned above have been overcome. This is effected by using
an auxiliary motor for starting purposes, thus keeping down
the starting current and at the same time producing sufficient
starting torque. The general arrangement of the connections in
a simple case is shown in Fig. 191, from which it will be seen that
at starting the full low tension pressure is made use of but the
stator windings of the induction motor used for starting are connected in series with the rotary slip rings. This arrangement has

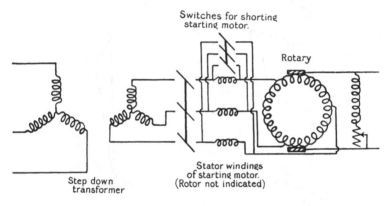

Fig. 191.

the effect of reducing the starting current passing through the
rotary armature to about 30 to 50 % of the full load current, a
value found in practice not to produce sufficient armature reaction
to prevent the residual magnetism of the field magnets of the
rotary always causing the polarity to build up in the correct
direction. The starting motor runs the rotary up to speed, the
field magnets excite themselves and the rotary pulls itself into
step; when this has occurred the stator windings of the starting
motor may be shorted by the switch shown and the rotary is
ready for paralleling on the D.C. side*.

* For a very complete account of self synchronising machines see a paper by
Dr Rosenburg, *J. I. E. E.*, Vol. 51, p. 220.

Pressure regulation on the direct current side of a rotary converter.
When rotaries are used for the purpose of converting alternating
to direct current for the supply of energy to lamps or motors, it is
clearly of the utmost importance that there should be convenient
means to hand for the purpose of keeping the direct pressure at
a suitable value as the load varies (it will usually be necessary
to increase the terminal pressure with increase of load).   The
apparently obvious way is to increase the excitation, either by a
shunt regulator or by a compound winding, but, as a matter of
fact, this may be said to produce no direct effect on the pressure

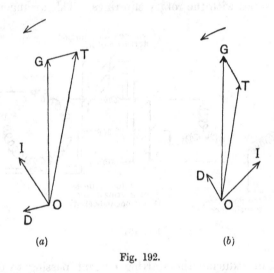

(a)   (b)

Fig. 192.

on the D.C. side.   It has been shown that the pressure on the D.C.
side depends directly on the pressure on the A.C. side and, since
this is settled by the supply voltage, both it and the pressure on
the D.C. side will remain constant (as far as direct action is concerned) no matter what the strength of the field may be.   What
actually occurs when the field strength is altered is that the phase
of the alternating current taken by the converter is altered (as
in a synchronous motor) and the altered armature reaction due
to the altered phase of the alternating current neutralises the
effect of the alteration in the field ampere turns.   Indirectly,
however, an increase in the excitation produces a small rise in the

alternating pressure applied to the rotary which will be accompanied by a corresponding rise in the pressure on the D.C. side. In Fig. 192 let OG be the pressure generated in the transformer secondary, this may be regarded as constant at all loads and is partly used in driving the current through the impedance of the circuit comprising the secondary winding of the transformer and the rotary armature and leads, and partly in balancing the pressure induced on the alternating current side of the converter.

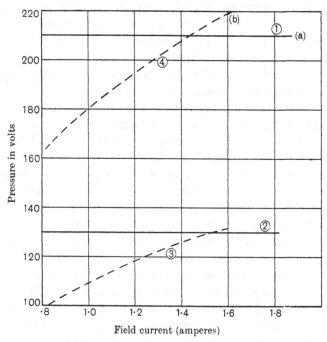

Field current (amperes)

(1) Pressure given by rotary on D.C. side (no reactance).
(4)   „   „   „   „   (added reactance).
(2) Pressure actually applied to rotary on A.C. side (no reactance).
(3)   „   „   „   „   „   (added reactance).
The pressure given by generator was kept constant throughout each test.

Fig. 193.

If the field of the rotary is over excited the current taken on the A.C. side will lead the pressure (as in the synchronous motor) and will be represented by OI in (a). The pressure drop in the circuit mentioned above will be represented by OD and on subtracting this from the generated pressure we arrive at the pressure, OT, applied to the rotary for conversion purposes which will practically settle the value of the pressure on the D.C. side. At (b) the corresponding vector diagram is shown when the rotary is under excited and is taking a lagging current, and it will be seen that, for a transformer secondary pressure equal to that in case (a), the pressure applied to the rotary is much less than before and will give rise to a less pressure on the direct current side.

It should be realised that the pressure variation produced on the D.C. side by varying the strength of the field current is due to the presence of reactance on the A.C. side, and this is well brought out in the curves shown in Fig. 193, which represent the result of varying the field current at no load on a converter directly supplied from a low tension alternator; in each case the alternator pressure was kept constant throughout the test but in (a) there was no reactance in the A.C. side of the converter other than the small amount due to the rotary armature, while in (b) a small choking coil was inserted in each line between the constant pressure alternator and the rotary. In practice the leakage reactance of the transformer serves in place of the choking coils used in the above test and, at constant load, a pressure variation of about 10 % on the D.C. side may be obtained by this method.

If a considerable variation of pressure is required on the D.C side it is usual to employ a booster placed between the low tension winding of the transformer and the rotary winding, thus allowing of the pressure applied to the slip rings being varied; the booster may be either of the static or synchronous type. The static booster has already been dealt with (see page 219), and a synchronous booster consists of an armature wound for the same number of phases and poles as the rotary, the current supplied to the rotary passing through the booster armature winding in its passage from the low tension transformer winding. The booster field is excited by direct current and by varying this a greater or less pressure may be added to or subtracted from the pressure supplied by the transformer. The booster is usually of the moving armature type and, since it is necessary that the phase of the booster pressure should correspond to that of the pressure supplied by the transformer, provision is made for the rotation of the field system through some small angle in order that this adjustment may be carried out.

It is worthy of note that it is possible to use the synchronous booster as the starting motor of a self-synchronising converter by temporarily using the field winding as the rotor winding of an induction motor, the armature windings of the booster of course acting as the stator windings on the starting motor; this method is used in practice by the Phoenix Dynamo Manufacturing Co.

Fig. 194.

A very interesting picture of a large two phase, 1200 k.w., rotary converter is shown in Fig. 194 (which has been prepared from a photograph kindly supplied by the British Westinghouse Manufacturing Co.). This rotary is provided with a starting motor, situated outside the near pedestal, and also with a synchronous booster of the revolving armature type. The novel and compact arrangement of the brush gear should be noted, this is necessary in order to convey the large currents to the machine without unduly increasing the size of the slip rings.

The Motor Converter.

The type of converting plant most recently introduced is the
la Cour motor converter which has been developed by Messrs
Bruce, Peebles and Co., who are the sole makers of this type of
machine for this country and the colonies, and to whom the author
is .indebted for information concerning the same and for the
diagrams from which Figs. 196, 197 and 198 were prepared.

It may perhaps be looked upon as a compromise between the
motor generator and the rotary and possesses to a large extent
the good points of both in addition to other qualities of its own.
In order to understand the mode of operation of the machine,
imagine an induction motor with a wound rotor and a rotary
converter (each being wound say for four poles) situated on the
same bed plate and mechanically coupled to each other; let
alternating current at 50 cycles per second be switched on to the
stator of the induction motor, the rotor being for the moment open
circuited, and let the set be driven by some external means. At
slow speeds the frequency of the pressure induced in the rotor of the
induction motor would be but little less than 50 cycles per second,
and the rotary would not excite itself owing to its slow speed.
As the speed of the combination rises the frequency of the pressure
induced in the rotor of the induction motor falls, and the other
machine excites itself, giving, at first, an alternating pressure whose
frequency is less than that of the pressure produced in the rotor
of the induction motor. As the speed of the combination rises
still further, the frequency of the pressure produced in the rotor
continues to fall, while the frequency of the pressure produced in
the rotary continues to rise and eventually (at a speed of 750
revolutions per minute in the case under consideration) the two
frequencies become equal, each being 25 cycles per second. If
the rotor and the rotary are wound for an equal number of phases
and so as to produce equal pressures under these conditions it is
clear that they may now be paralleled, due care being taken to
connect the corresponding lines of the two machines to each other
with the pressures in opposition; no current will, for the moment,
flow between the two machines, owing to the equality in magnitude,
and phase opposition, of the pressures involved. The state of

affairs in the local circuit comprising the rotor of the induction motor and the rotary may be indicated as at (a) in Fig. 195, where OE_1 represents the rotor pressure and OE_2 the rotary pressure. Now imagine the external driving force to be removed, since there is no rotor current there can be no electrical driving force, and the set will commence to fall in speed with the immediate tendency to cause a rise in the rotor frequency and a fall in the rotary frequency. When the machine has fallen a fraction of a revolution behind the position corresponding to a speed of 750 revolutions per minute, the vector diagram for the internal circuit of the machine

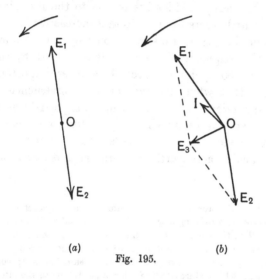

(a)   (b)
Fig. 195.

will be as shown at Fig. 195 (b). There is now a resultant pressure OE_3 in the local circuit comprising the rotor and the rotary armature and this will send a current OI which will lag considerably behind the resultant pressure owing to the relatively large value of the reactance compared with the resistance of the circuit in question.

If we examine this current in conjunction with the rotor pressure OE_1 we see that it is in much the same phase relationship as would be the rotor current in an ordinary induction motor and we deduce that the induction motor side is taking power from the mains and developing torque.

Again, if we examine the action of the current in conjunction with the rotary pressure, we see that the phase relationship between OI and OE_2 is such as would exist in an ordinary rotary and we deduce that this part of the machine is motoring from the A.C. side and is, of course, prepared to supply current to an external circuit on the D.C. side. It is necessary to clearly grasp the fact that the slowing down referred to above would only be of a momentary nature, the combination simply drops back a little, relative to the position of true phase opposition, until sufficient power is taken from the mains to drive the set, and then continues to run synchronously. If load is put on to the D.C. side the set drops back a little more in order to take increased power from the A.C. side and then again runs synchronously. It should also be noted that the frequency of the currents dealt with by the rotary is 25 and not 50 cycles per second—a most important point. We see then that what, for the sake of convenience, may be called the induction motor side, is operating partly as a motor (really as a species of synchronous motor), and partly as a pressure, frequency, and may be a phase transformer, while the other machine acts partly as a rotary and partly as a D.C. generator*.

---

* In an induction motor the rotor and stator currents may each be looked upon as tending to produce a rotating magnetic field (the actual field being the resultant of the two), further, the speeds of rotation of these two component fields, in space, are in every case equal.   The speed of rotation of the stator component field is produced simply by the alternations of the supply pressure, but the speed of rotation of the component field produced by the rotor is partly due to the alternations of the rotor current and partly due to the motion of the rotor conductors.   In the ordinary induction motor both tend to give a forward motion to the rotor component field.   If we excite the rotor with direct current from an external source, the frequency of rotation of the rotor field is simply due to the motion of the conductors and the machine will run synchronously as is the case with the ordinary synchronous motor.   If we supplied the rotor with alternating current from an external source such that, due to the alternations of the current, the rotor field tended to rotate backwards, the rotor itself would run up to a speed above synchronism so that the resultant speed of the rotor component field was equal to that of the stator field. In the case of the induction motor side of the motor converter the alternations of the rotor current are such as to cause the rotor component field to rotate with a considerable speed in a forward direction, with the result, that the rotor conductors need to rotate only comparatively slowly in order that the speed of the rotor component field in space may be equal to that of the stator field.

In practice it is permissible to wind the A.C. stator for high tension current, thus doing away with the necessity of a separate transformer (compare with the case of a rotary converter); the number of phases for which the stator is wound is usually three. The rotor and rotary are usually wound for six or twelve phases and, since both are revolving at the same rate, it is unnecessary to use slip rings between the two, the connections being made directly through the hollow shaft. It is perhaps unnecessary to state that the method of starting described above was only suggested in order to arrive at a simple explanation of the action of the machine, the method used in practice is far simpler.

The A.C. and D.C. machines in the example quoted above had equal numbers of poles, but this is not essential and the two machines may have different numbers of poles if desired, the synchronous speed of the set always being such that the frequencies of the rotor and rotary pressures are equal. Thus, let the number of poles of the alternating current machine be p_1, and the number of poles of the direct current machine (*i.e.* the rotary) be p_2, the frequency of the supply current being f cycles per second and the speed of the set R.P.M. revolutions per minute.

The frequency of the pressure produced in the rotary will be

$$\frac{p_2}{2} \times \frac{\text{R.P.M.}}{60} \text{ cycles per second,}$$

and the frequency of that produced in the rotor will be

$$\left(\frac{f \times 60}{\frac{p_1}{2}} - \text{R.P.M.} \right) \times \frac{p_1}{2 \times 60} = f - \frac{\text{R.P.M.} \times p_1}{120} \text{ cycles per second.}$$

These must be equal and we therefore have

$$\frac{p_2}{2} \times \frac{\text{R.P.M.}}{60} = f - \frac{\text{R.P.M.} \times p_1}{120},$$

or

$$\text{R.P.M.} \left(\frac{p_2}{120} + \frac{p_1}{120} \right) = f;$$

$$\therefore \text{R.P.M.} = \frac{f \times 120}{p_2 + p_1}.$$

That is to say the set revolves at a synchronous speed corresponding to the sum of the number of poles in the two machines.

Starting of motor converters.

(1) From the alternating current side.

The general arrangements (including those for starting from the A.C. side) of a motor converter set are shown in Fig. 196, from which it will be seen that a star connected rotor is employed, the star point being formed, when running, with the help of the short circuiting gear indicated. For starting, three phases are brought out to slip rings provided with brushes to which a standard three phase starting resistance can be connected. When the high tension alternating current is switched

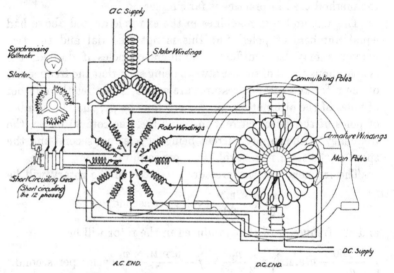

Fig. 196.

on to the stator the set starts up as an induction motor on three phases, the rotary winding being used for the moment for connection purposes only. As the machine approaches the synchronous speed appropriate to the converter the rotary excites itself, and, just below synchronous speed, we shall have the rotor and rotary pressures alternately helping and opposing each other, thus causing large variations in the magnitude of the current in the starting resistance and also in the readings of the voltmeter connected thereto. If the starting resistance is adjusted to a suitable value the machine then pulls itself into step and this is indicated by the voltmeter needle ceasing to fluctuate violently and taking up a small steady reading. The device for connecting together all the inner ends of the phases is then pushed home, the brushes lifted from the rings (unless use is to be made of the machine for static balancing) and the machine is ready for paralleling on the D.C. side.

(2) From the direct current side.

The machine may also be started by using an ordinary motor starter on the D.C. side (notice that some of the armature starting current will be shunted through the rotor windings), the set being paralleled with the A.C. high tension line pressure after full speed has been reached.

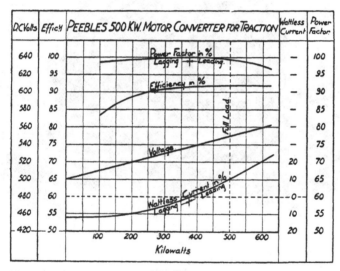

Fig. 197.

Effect of varying the direct current excitation of a motor converter. When the direct current excitation is increased two effects are observed, in the first place the pressure on the D.C. side rises and this may be looked on as an analogous effect to that which occurs when the excitation of a rotary is increased; the effect is usually of greater magnitude than that occurring in the rotary. There is also an action on the phase of the current taken by the stator when the excitation is increased, again analogous to that occurring

19—2

in the rotary and synchronous motor, the power factor being low, due to lagging currents, when the excitation is low, while by increasing the excitation the power factor rises and becomes unity, and for still higher excitations the current taken by the stator may be a leading one. For lighting purposes the motor converter is usually shunt wound, the required pressure regulation being obtained by a shunt regulator of the ordinary type, and the set is designed to operate very nearly at unity power factor at all except small loads. For traction purposes an over compounded

Fig. 198.

machine is usually employed. The changes in field strength brought about in obtaining pressure regulation also affect to some extent the power factor on the A.C. side, this usually being below unity (current lagging) at small loads, unity at some intermediate load, and again below unity (current leading) as full load is reached; the departure from unity power factor is never very serious in amount.

The performance curves of a 500 K.W. set which is over compounded for traction purposes are shown in Fig. 197, in which the above points are well brought out.

A point in connection with the motor converter, on which great stress is laid by the makers, is the great synchronising force which occurs in the machine and which results in the machine keeping in step even when the applied pressure and frequency temporarily fall, as may happen in the event of a short on the line, and this claim is borne out by many users of such machines. The efficiency of the motor converter is comparatively good at the smaller loads but is lower than that of a rotary, working under the same conditions, at or near full load. A general view of a 1500 K.W., three bearing type of motor converter is shown in Fig. 198, the slip rings and short circuiting gear being clearly seen.

The static balancer for direct current three wire circuits. This is a device that is coming into use for balancing purposes in

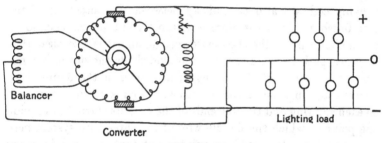

Fig. 199.

sub-stations in which rotary converters or motor converters are employed for converting from high tension alternating current to low tension direct current for lighting purposes.

As is well known it is customary in direct current lighting circuits to supply energy on the three wire system with a pressure of from 400 to 500 volts between the outers, the corresponding pressure between either outer and the middle wire being from 200 to 250 volts. By using this system the pressure supplied to the consumer is kept within legal, and also within lamp, requirements, yet the economy of copper due to distribution at double the pressure is, to a large extent, secured. It is very essential, however, to provide some means to keep the potential of the middle wire midway between that of the outers and, at the same time, provide a path back to the generator for the out-of-balance current.

In most direct current schemes it is usual to employ rotary balancers, but, under certain circumstances which are mentioned below, it is possible to use a static balancer which is cheaper and simpler in operation.

Consider a direct current generator which is provided, in addition to the usual commutator, with two slip rings connected to the winding in such a manner as to produce single phase current; across the slip rings let a choker be connected as shown in Fig. 199. The pressure applied to the choking coil will of course be an alternating one, but it is clear that at every instant the potential of the middle point of the choker will be midway between the potentials of the two slip rings and, since at every instant the positions of the slip ring connections will be symmetrical with respect to the brushes, the pressure of the middle point of the choking coil will also be midway between the potentials of the two brushes. It is therefore a suitable point for connection to the middle wire of the three wire system, and it is also clear that a return path to the armature is provided for the out-of-balance current. Instead of the arrangement shown in the figure three or more slip rings may be used, one end of a choker being connected to each slip ring and the free ends of the chokers starred to provide the point to which the middle wire of the three wire system may be connected. It should be pointed out that it is not in every case necessary to use external chokers, thus, in a motor converter, the middle wire may be connected to the slip rings used for starting purposes, the starred windings of the rotor of the A.C. side of the motor converter then being used for the choking coils. Similarly, with rotary converters, if the low tension side of the transformers supplying current to the rotary be starred, the middle wire may be brought to this point and, in addition to their ordinary function, the secondary windings of the transformers will also serve as the choking coils for static balancing. The automatic balancing effected by this method is not perfect, a slight difference of pressure developing between the two sides of the D.C. system as the out-of-balance current increases, but this may be made of the order of about 1 % for the largest proportion of out-of-balance current usually occurring.

The Mercury Arc Rectifier.

A type of converter which has of late come into considerable use for small powers and which is quite likely to have greater application in the future, perhaps for much larger powers, is the electric valve. As is well known it is possible to construct pieces of apparatus which allow of the flow of current in one direction much more readily than in the opposite direction, thus, if an electrolytic cell is made having plates of aluminium and lead respectively, and with a neutral solution of ammonium phosphate as the electrolyte, the phenomenon is well exhibited, a considerably higher pressure being required to drive a current through the liquid from the aluminium to the lead than is required to send current in the opposite direction. The phenomenon is, in this case, due to the formation of a non-conducting film on the surface of the aluminium, and it is to be noted that this film only stops the flow of current from the aluminium to the lead practically completely up to a certain pressure (the best working pressure is of the order of 140 volts) above which the leakage through the cell increases rapidly. The phenomenon is also exhibited in the ordinary mercury vapour lamp which permits current to flow from the iron or carbon electrode to the mercury electrode much more readily than in the opposite direction. It is perhaps hardly necessary to mention that these devices are only used for the conversion of alternating to direct current and are not available for the conversion of direct to alternating current.

When arc lamps are to be used for projection purposes a much better and steadier light can be obtained by using direct than by using alternating current, and if the supply is in the form of alternating current the mercury arc rectifier offers great possibilities as a means of converting the current applied to the lamp into a direct one. An arrangement suitable for this purpose is supplied by the Westinghouse, Cooper, Hewitt Co., to whom the writer is indebted for information concerning the matter and for the diagram of connections shown in Fig. 200.

This figure represents the connections of a single phase converter supplied from a step down transformer having a secondary wound so as to produce an alternating pressure which will convert into the required magnitude of direct pressure for application to the arc lamp. When a single arc lamp is used the pressure on the D.C. side is approximately one-third of the pressure between the extremities of the low tension winding of the transformer. If it is desired to vary the pressure on the D.C. side a number of regulating turns, controlled by a suitable switch, should be provided on the transformer. The bulb has two anodes marked A and one cathode K, the current always flowing to the latter but coming alternately from each of the anodes, advantage is thus taken of each half wave of the alternating current with resulting steadiness in the pressure and current on the D.C. side. The large bulb Z at the top of the tube is arranged to receive the vapour of the mercury, condense it, and then return it to the cathode; it also serves as radiating surface for the dissipation of the heat resulting from the condensation, thus assisting to keep the bulb cool.

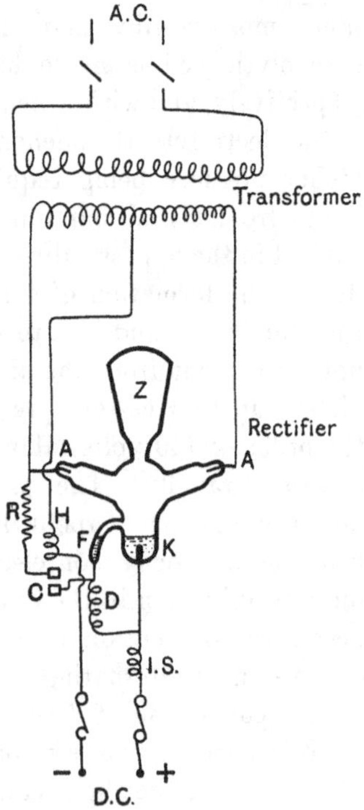

Fig. 200.

The tube is not self starting without the help of some external device since, before the current can flow, a stream of conducting vapour must be produced (*i.e.* ionisation must be taking place at the cathode).

The actual process of starting is as follows: the double pole A.C. switch having been closed and the carbon electrodes of the D.C. arc brought into contact, a path is provided for the flow of current through one half of the low tension side of the transformer, the starting contacts C, and the coil D which causes the tube to tilt until mercury flows from the cathode to the special starting electrode F, this short circuits the tilting coil D and allows the tube to fall back to its original position thus starting a short arc between F and K. This short arc provides the necessary conducting vapour for the passage of arcs between the cathode and one or other of the two anodes. As soon as this is effected, direct current flows in the arc lamp circuit and the starting circuit is broken by the coil H operated by the arc lamp current. The effect of the self induction IS is threefold, it prevents a large rush of current, which might result in blown fuses, when the carbons are brought into contact, it steadies the direct current, and also prevents the tube shutting down when the alternating current passes through its zero value.

In some applications of the converter, as when charging accumulators, it is necessary to start up the converter on an auxiliary resistance, the battery being afterwards substituted for it by means of a two way switch. This resistance may be permanently retained in parallel with the load if the latter is of an intermittent character. The glass tube converter has, at present, a maximum capacity of 40 amperes on the D.C. side, and, when larger currents are required, two or more bulbs may be worked in parallel, but progress is being made with metal bulb converters which present possibilities of, and in fact have actually been made for, much larger outputs. When the supply is three phase the low tension side of the step down transformer is starred, the cathode being connected to the neutral point and three anodes provided. One of the greatest troubles to be feared with this type of converter is arcing over directly from one anode to another and this risk is minimised by separating the anodes as far as possible and in some cases by making the path from one anode to another of a tortuous nature. The bulbs can be made for use on circuits having very different pressures extending up to several thousands of volts. For pressures on the D.C. side of 200 volts and upwards efficiencies of say 90 % may be anticipated, the power factor on the A.C. side being of the order of ·9 except when considerable inductance is made use of in order to steady the direct current in which case lower values are obtained.

The current obtained from such a converter is not a steady one and in fact fluctuates between considerable limits though always in the same direction.

CHAPTER XI

SWITCHGEAR AND PROTECTIVE APPLIANCES;
HIGH TENSION TRANSMISSION

A typical diagrammatic arrangement of the generator switch-gear in a high tension generating station is shown in Fig. 201, in which T represents the trifurcating box in which the three core cable (if such is used) coming from the generator is attached to

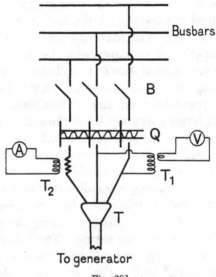

Fig. 201.

the separate tails leading to the bus bars, T_2 the current trans-former for use with the ammeter, and T_1 the pressure transformer for use with the voltmeter. The main oil break switch is shown at Q, and, since it is necessary to make this "dead" for inspection, cleaning and repair, isolating switches are placed at B. Instead of a simple isolating switch as shown, a selector switch may be substituted in order that any generator may be placed on to either of two sets of bus bars, but in any case it is not intended that current should be broken at such a device and it is not designed for that purpose.

In addition to the functions of T_2 and T_1 mentioned above, they may also be employed in connection with the auto tripping or indicating devices for the protection of generators, though

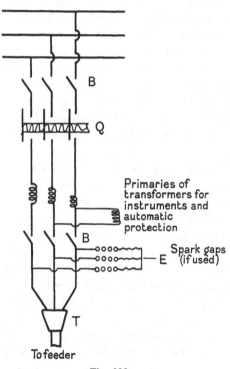

Fig. 202.

separate transformers are often used for this purpose. It is clearly desirable to limit the number of instrument transformers as much as possible or they will become very numerous and each is a potential source of weakness.

The general arrangements for a high tension feeder are as shown in Fig. 202, and it will be seen that they are very much the same as those used in connection with generators. The most important difference between the two diagrams is that, in the case of the feeder, isolating switches are shown on each side of the main oil break switch, this is necessary because a feeder is usually liable to be made alive from either end.

Perhaps the most important link in the gear shown is the main switch which, in modern large high pressure stations, may be called upon to perform a very difficult duty. The switches used must be capable of carrying the full load current of the feeder or generator continuously without over heating and must, in addition, be capable of breaking a current very much in excess of the full load current in a satisfactory manner, since in emergencies they may often be called upon to effect this operation. Modern switches for use on high tension alternating current circuits are almost invariably oil immersed, the quantity of oil used being such as will cover the contacts to a considerable depth. In three phase switches it is also customary to provide three separate oil tanks for the three phases, though a single tank, provided with suitable partitions, may be used for moderate currents and moderate pressures. Such oil immersed high tension switches may be operated directly or by some form of remote control, the latter method being of particular service in the case of switches suitable for dealing with large powers, under which circumstances the moving parts will be very heavy. When remote control is employed the switches may be operated mechanically, by the use of linkwork, or electrically, in which case the solenoidal type of control is most common. A high tension switch as made by Messrs Ferranti, Ltd. (to whom the writer is indebted for several diagrams and much information concerning their switches), and which is suitable for directly mounting behind the switch board panel, is shown in Fig. 203. The arrangement of the contacts of one phase is shown in the upper portion of the figure where the parts C represent the fixed contacts, current being conveyed to these, from the cables forming the circuit to be controlled, by connections passing through the porcelain bushes P which are secured to the cast iron cover of the switch K. When the switch is closed the two fixed contacts are bridged across by the laminated bridge piece B which is carried by a specially prepared teak rod T. The upper end of the rod is secured to (in

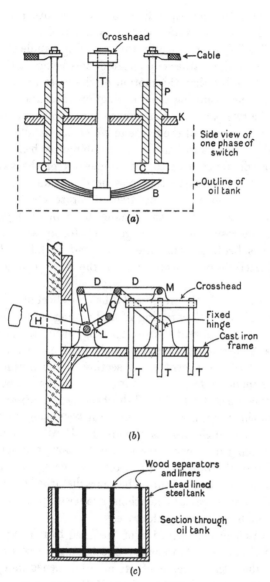

Fig. 203. Sketches of essential parts of three phase oil immersed switch
(not to scale).

common with the corresponding rods of the other phases) the forged iron cross head whose motion is actuated by the operating handle. The link gear, whereby the operating handle is connected to the cross head, is rather complicated, this being due to the fact that it is desirable that the motion of the cross head be exactly vertical thus necessitating the introduction of a paralleling motion between the two parts. The arrangement of this motion is shown in the central portion of the figure in which the handle H is rigidly connected to the short bar B, the combination being hinged at L about a spindle secured to the framework of the switch; the long arm D is pivoted to the cross head and also to the small link K, the latter link being free to rotate about the pin L. When the switch is open the handle H is in its highest position with the result that the toggle joint (composed of the links B and C) is bent, allowing the cross head to fall and carrying the laminated bridge pieces free of the fixed contacts of the switches.

When the handle H is depressed it has the effect of straightening out the toggle joint thus raising M and closing the switch.

The switch is prevented from opening, owing to the weight of the parts, by arranging that the toggle joint has a final position just beyond the dead centre. A section of the containing case, which is supported from the casting forming the cover, is shown at (c); it is composed of lead lined sheet steel and for safety is further lined with three ply maple, this material also being used between the phases if a single tank is employed. It will be noticed that the wood lining does not come into continuous contact with the metal sides of the tank, a layer of oil intervening; by this means, even if the general body of the oil becomes disturbed by the opening of the switch under severe conditions, a layer of quiet oil is secured between the live parts and the case.

The switch just described is of the fixed handle type but it is more usual and better to employ one having a free handle, that is one in which the switch contacts may be opened independently of the motion of the handle. Thus, if the switch is closed under faulty conditions, it can immediately re-open (if automatic protection is used) even if the handle is firmly grasped, which is not the case with a fixed handle switch.

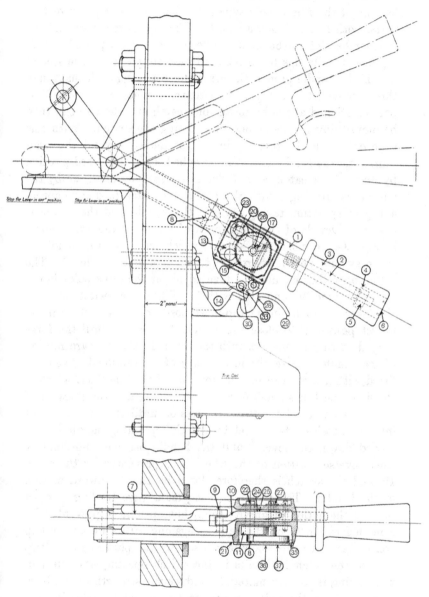

Fig. 204.

The above pattern of switch may readily be modified so as to become of the free handle type if the handle is composed of two parts, both being hinged about *L*. The first part corresponds to the member *HB* in the fixed handle type but is not provided with a proper handle this being replaced by a simple lever, the second part is the handle proper, the arrangements being such that when this is moved to its top-most position it engages with the first part mentioned above by means of a catch and the two can then be moved downwards as one thus closing the switch. With this type of switch the toggle joint is not moved beyond the dead centre and the handle must therefore be held in the closed position by means of a catch, and if this is released by hand, or by the automatic tripping device, the part *H* flies back (owing to spring and gravity action taking place in the switch) and the switch is opened. Details of the mechanism of the free handle are shown in Fig. 204, in which the lever 7, which served as the handle in the previous type, is short and terminates in the roller 8. The actual operating handle 4 is pivoted about the same point but is not permanently connected with 7. When the switch is open the lever 7 will be in its top-most position but 4 will be in its lowest position; to close the switch, 4 is raised until the fork-shaped casting 10 engages with the roller, further upward motion of the handle then results in the back of the casting 10 (which is fitted with a rack) causing the rotation of the wheel 17, which is the first wheel of a small enclosed gear train, and the rotation of the last member of the train, 13, goes on until the catch 15 slips into the notch on the wheel 13. If the operating handle is now moved down, the lever 7 and the handle become co-ordinated, since reverse rotation of the wheel 13 is prevented by the catch 15, and the switch is therefore closed by the downward motion of the handle. The switch is of course always tending to open (due to the action of gravity, etc.) but opening is prevented by a further catch, 28, on the handle engaging with a lip on the trip coil casing. A slight blow, given say by the plunger of the trip coil, to the catch 15 results in the switch opening (the effect of the gearing is to very materially reduce the strength of the blow required), and the switch may also be opened by releasing the catch 28 by the lever 29. A reproduction from a photograph of a three phase high tension switch with free handle and suitable for mounting directly behind the switch board panel is shown in Fig. 205.

The type of switch described above is directly applicable in cases in which mechanical remote control is required, the operating handle being connected to the link work of the switch by a series of rods and bell crank levers as indicated in Fig. 206.

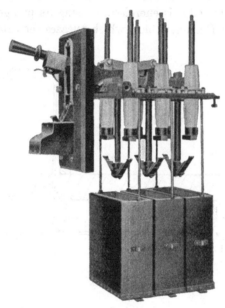

Fig. 205.

The general arrangements of a high tension oil immersed switch (also made by Messrs Ferranti, Ltd., to whom the writer is indebted for the photograph of the switch and for information concerning the same) which is suitable for remote electrical solenoidal control are indicated in Fig. 207. The main solenoid for closing the switch is enclosed within the casing A; when this solenoid is energised it causes the switch to close by sucking up a core composed of a number of concentric slotted iron tubes. Since considerable power is needed to energise this solenoid, arrangements are made so that it is only necessary to keep current flowing in the coil for a brief instant, the switch being then held in the closed position by means of a catch. This is effected by means of a non-magnetic rod B, attached to the top of the core, which elevates the top of the link work C until a notch on the horizontal rod D is caught by the spindle E; this occurs just as the contacts reach the closed position. The power required to close a large switch of this type may be as high as 10 K.W. but it will only be required for a short time of the order of one second. When it is required to open the switch the trip coil F is energised, causing its plunger to rise and strike the rod D a smart blow which releases the catch and allows

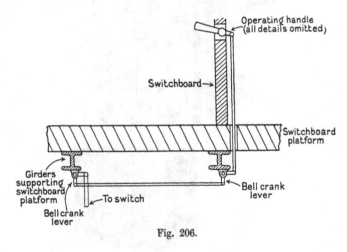

Fig. 206.

the switch contacts to open owing to the combined effects of gravity and spring action. G is a small controller, operated from the top link work, which opens the circuit of the tripping coil as the main switch opens, thus preventing arcing at the contacts of the relay or press button by means of which the trip coil was energised; it also has contacts to operate the lamps on the switch board which are used to indicate whether the main switch is open or closed. It is usual to attach springs to the moving portion of the switch in such a way as to assist the large solenoid in lifting the moving contacts when the latter commence to rise, and when the contacts are nearly closed the springs are arranged to oppose the solenoid thus taking off a certain amount of the mechanical shock that ensues at the moment of closure. It is clear that these springs will assist gravity at the moment the switch starts to open and when the switch contacts approach the full open position the

Fig. 207.

springs will again tend to diminish the mechanical shock. The operating solenoids are usually arranged for working with direct current at a pressure of from 50 to 200 volts.

20—2

It is very necessary that the control gear of such a switch as is described above should be certain in action and should not permit of any hesitancy on the part of the operator affecting the operation of the switch. The control drum used by Messrs Ferranti and Co. consists of two parts, the drum proper mounted at the back of the board, and the loose handle mounted at the front of the board, the two being connected by a suitable spring. To close the switch the control handle is turned (it can only turn in one direction, reverse motion being prevented by a ratchet and pawl) until it is caught by a catch, this has the effect of extending the spring connecting the handle and the drum, the latter being prevented from rotating by means of a catch which can be released when necessary. The drum is then in a position ready for use

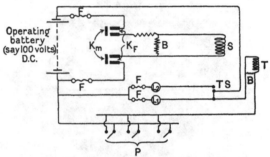

F. Fuses. K_F. Fixed contacts of main control drum. K_m. Moving contacts of main control drum. *B*. Buffer resistances. *S*. Solenoid for closing switch. *T*. Trip coil. L_1. Lamp to indicate when main switch is closed. L_2. Lamp to indicate when main switch is open. *TS*. Two way switch operated by motion of main switch. *P*. Push button and auto trip coil contacts.

Fig. 208.

and when the catch is released, by pressing the button provided for the purpose, the drum rotates at a speed determined by the motion of a pendulum and closes the circuit of the main solenoid for a time which is ample for the main switch to close; further motion then results in the circuit of the main solenoid of the switch being opened, a suitable discharge resistance being provided to prevent excessive sparking at the contacts on the drum. A second press button is provided for opening the switch and this is simply arranged to close the circuit of the trip coil shown on the switch and, as stated before, the operating current for this purpose is broken on the small drum controller situated on the main switch in order to prevent damage to the press button contacts due to arcing. It is usual to provide mechanical means of opening and closing the switch for use in case of failure of the electrical method. A complete scheme of connections for the operating circuits is given in Fig. 208; this diagram will be self explanatory in view of the information already given.

Isolating switch. An isolating switch is usually a slow break switch without a handle, being opened, when desired, by means of a hook attached to a wooden rod. It is essential that the contacts shall be capable of carrying the full current continuously and that all live parts of the switch be insulated sufficiently for the pressures to be dealt with.

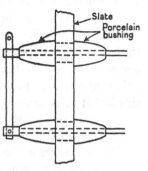

Fig. 209. Isolating switch.

Relays for use with switches controlling generators. The question of the automatic protection to be afforded to generators has been the subject of much discussion; in the first place it is clear that when generators are working in parallel it is inadvisable to use overload protection since if, owing to over load, one generator switch is tripped, additional load will be thrown on to the other generators and these are likely to be tripped in turn thus causing a complete shut down. What is really necessary is to have a means of cutting off a generator should it take power from the bus bars instead of supplying power to the bus bars, and this state of affairs is most likely to happen as the result of a short or earth occurring in the generator winding or in the event of a failing prime mover. The problem is complicated owing to the fact that with alternating current we have no definite direction of current corresponding to power being supplied to the bus bars, and no definite reverse direction when power is being taken from the bus bars, as is the case with direct current. It is simply the relative phase of the machine pressure and current which acts as the criterion as to whether the machine is generating or motoring. Thus, in Fig. 210, if the vector representing the current in the machine is in the region marked (*a*), the machine would be supplying power to the bus bars, but if in the region marked (*b*), the machine would be taking power from the bus bars. One of the most difficult problems to deal with in an attempt to provide automatic protection for generators is the case of a failing generator field. In such a case the machine may still be supplying power* to the bus bars but it will be at a very low power factor and the set will serve little useful purpose, simply loading up the other generators with wattless cur-

rent, so that it will be best that such a generator be removed from the mains. An independent tripping device, actuated by the direct exciting current, may be used to effect this but it is possible to adjust an ordinary reverse power relay to effect the same end†.

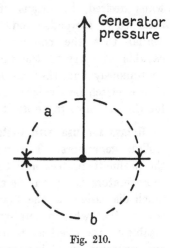

Fig. 210.

Reverse power or discriminating relays have been constructed on the solenoidal and on the watt hour meter principles.

In the latter case an ordinary watt hour meter movement of the induction type (see page 140) may be connected to the generator leads through suitable current and pressure transformers: as long as the generator is giving power to the bus bars the tendency will be for the movable disc to revolve in one direction and matters may be arranged so that this motion is prevented by means of a stop: if the generator is taking power from the bus bars the disc will tend to rotate in the opposite direction, and in this case the motion may be allowed to close the trip coil circuit of the main switch.

The arrangement used for this purpose by Messrs Ferranti, Ltd., is shown in Fig. 211 from which it is seen that so long as power is

---

* See paper by Dr C. C. Garrard, *J. I. E. E.*, Vol. 41, p. 597.
† See paper by Mr Andrews, *J. I. E. E.*, Vol. 34, p. 438.

being given to the bus bars the disc of the watt hour meter
movement tends to rotate clockwise and in fact holds the
contacts of the relay open, but as soon as reverse power causes
the disc to rotate in the counter clockwise direction a weight is
lifted and the contact of the trip circuit of the main switch is
closed owing to the lever C being knocked over by the striker A.
The amount of reverse power at which the relay operates is determined by the magnitude of the weight and a time element can,

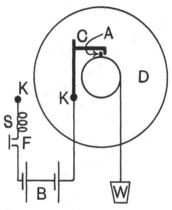

S. Tripping solenoid on main switch. K. Contacts in trip circuit which are
closed when lever C is knocked over by striker A. F. Auxiliary contacts in trip
circuit which are opened when the main switch is opened thus avoiding sparking
at K. B. Trip circuit battery. W. Weight to fix reverse power at which
relay operates. D. Copper or aluminium disc of wattmeter movement. (Note
that the current and pressure coils of the wattmeter are not shown in the figure.)

Fig. 211.

if desired, be introduced by damping the motion of the disc by
means of a permanent magnet.

If a time element is introduced, however, there is a risk of the
bus bar pressure falling considerably and this would cause the
sense of direction of the relay, which is given by the pressure coil,
to be considerably weakened or possibly practically lost altogether;
the relay might then fail to work or at any rate a much larger
current (one representing reverse power) would be required to
operate the relay.

Relays working on the Merz Price system may also be used to trip the generator in the event of an internal short or earth (see page 315).

In some cases reverse power relays are not used to trip the main switch directly but are connected so as to light up a signal lamp, thus indicating the trouble to the switch board operator who can then open the main switch by the usual hand tripping device.

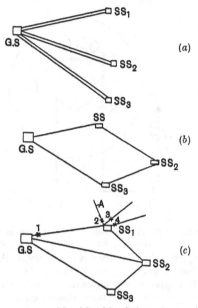

G.S. Generating station. *SS*₁, *SS*₂, *SS*₃. Sub-stations. A single line indicates a three phase feeder.

Fig. 212.

Arrangement of feeders, etc. When a generating station is arranged to feed a number of sub-stations (whether these are the property of the supply company and used for a general supply to small consumers, or whether they are situated on the premises of a large consumer and used entirely for the supply of that consumer) it is of the utmost importance to arrange a duplicate transmission line so that in the event of a breakdown of a section of a cable there shall still be a continuity of supply. This may be effected in several ways, perhaps the most obvious one being to run duplicate feeders which, under normal conditions, may be operated in parallel or singly. From the point of view of the continuity of the supply it would obviously be preferable not to run them by the same route but at some considerable distance from each other for the greater part of their length, though economic reasons will be against this course. Another way would be to make use of a ring main, as indicated in Fig. 212 (b), from which it will be seen that it is possible to supply a certain consumer either way round the ring. If the whole of the ring is sound a break may be made at any desired point by means of an inter-connecting switch. Finally, it is possible to make use of a combination of single feeders and inter-connectors as in Fig. 212 (c), the inter-connectors in normal operation may or may not be used, as is considered desirable. With such a possible variety in the feeder arrangements, only the simplest cases being mentioned above, the question of the automatic protection of feeders becomes complicated and assumes many forms. In general it may be stated that it is desirable to cut off a section of the feeding cable in the event of very excessive current and in the event of a fault (either of the nature of an earth or of a short) in the section concerned.

Over load protection of feeders*. This can readily be effected by the use of a suitable relay, of which there are many types available, arranged to trip the switch either mechanically or electrically. Such relays are commonly placed on feeders where they leave the generating station and also on sub-feeders leaving sub-stations, and it is desirable to place a relay (capable of independently tripping the switch) on each line. If a combined relay is used so that a reasonable over load occurring in each line simultaneously operates the relay, a very excessive over load will be necessary should it occur in one or two lines only. Overload relays, unless they are very carefully arranged, are apt to cut off a greater section of feeder or main than is really necessary thus causing inconvenience to an unduly large number of consumers: thus, in Fig. 212 (c), in which feeders and sub-feeders are shown protected by over load relays at the points 1, 2, 3, etc., if over load occurs

* See paper by Dr C. C. Garrard, *J. I. E. E.*, Vol. 41, p. 588.

(perhaps due to a short) at the point A, matters should be arranged so that the relay at 2 operates thus cutting off the faulty section, but the relay at 1 should not operate since by so doing it would cut off a considerable amount of perfectly sound main. It might be supposed that this end would be attained by setting the relay at 2 to operate with a smaller current than the relay at 1, but with the very large currents likely to occur under short circuit conditions it is very probable that both would be operated simultaneously.

A better way is to instal relays with a time element, which may be of the inverse or of the fixed type (preferably of the latter since an inverse time element means almost instantaneous operation with very excessive currents). If the time elements are fixed and if that on 1 is made of longer dura-tion than that on 2 then, under the conditions stated above, the switch at 2 will be operated but that at 1 will not open and this is the end desired.

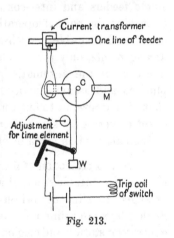

Fig. 213.

A simple and interesting type of over load relay as made by Messrs Ferranti is shown in Fig. 213, it consists of a shielded pole motor (exactly similar in principle to the ammeter working on the same lines described on page 111) which is operated, in high tension systems, from the secondary of a suitable current transformer though it may of course be wound to take the main current if desired. The motion of the copper disc C is retarded by the eddy currents produced by the permanent magnet M, and when the disc revolves it winds up the weight W and eventually closes the tripping circuit of the switch by means of the bell crank lever D.

The particular over load at which the relay operates is settled by the magnitude of the weight, and the time that elapses between the moment at which the over load comes on and the moment at which the relay operates is (to some extent) settled by the distance through which the weight has to be lifted and means are provided for adjusting this. Should the over load go off before the trip circuit is closed, the relay will reset itself and this also happens as soon as the switch has been tripped.

The Merz Price system of protection. This is an arrangement whose essential function is to cut off a feeder, inter-connector, etc., in the event of an earth or short occurring thereon, but it may also be utilised for over load protection in certain cases if desired. It depends upon the fact that in a certain section of cable the current leaving will, so long as the section under consideration is sound, be equal to the current entering, except perhaps for some small difference due to the capacity of the cable. In Fig. 214 let a, b and c be the three cores of a three phase cable, each core being provided with a similar type of current transformer at each end.

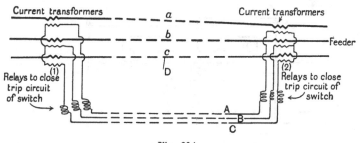

Fig. 214.

In the figure three ends of the secondaries of the current transformers at one end of the feeder are starred, the remaining ends being connected to the pilot wires A, B and C which are also connected to the secondaries of the current transformers at the far end of the feeder in a precisely similar manner. In the case shown it is intended that in the normal operation of the system the pressures in the secondaries at one end shall just balance the pressures in the secondaries of the corresponding transformers at the other end, thus, so long as the main cables are quite sound, there will be no current flowing in the pilot wires. Now suppose that an earth occurs at the point D (the neutral point of the system being supposed to be earth connected through a suitable limiting resistance), then there will be a greater current flowing through the primary of current transformer (1) than through the primary of current transformer (2), and a greater pressure will be induced in the secondary of current transformer (1) than in the secondary of current transformer (2) with the result that a current will flow along the pilot wire C^*. The current thus produced in the pilot wire will energise the relay and so close the trip coil of the switch. The relays can be made very sensitive and only a small current is required in the pilot wires and this in turn necessitates but a small out-of-balance current, caused by the fault, to operate the device. It is possible therefore to earth connect the neutral point of the system through quite a high resistance and the line disturbance caused by an earth sufficiently large to trip the switch is very small. If the neutral point of the system is not earthed a capacity current will flow to earth through the fault and thence to the other two cores, not only of the same feeder but also of the other

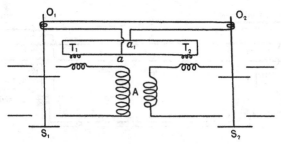

A. Main power transformer. S_1, S_2. Oil break switches. T_1, T_2. Current transformers for Merz Price Gear. O_1, O_2. Trip coils of main switches or coils of relays arranged to trip the main switches.

Fig. 215.

feeders, and in large systems this capacity current may be large enough to cause the relays to operate. It will also be clear that the relays will operate in the event of a short circuit between two lines.

It will be realised that the system as outlined above does not afford any over load protection but this can be provided, if desired, by the use of a separate relay of a type already described. The Merz Price system can also be used for the protection of transformers and alternators against earths and shorts and it is particularly useful in the case of inter-connectors; in the latter case, power may be transferred either way along the main while it is healthy, and thus reverse energy relays are quite useless. An

* Completing its path, of course, through the other pilot wire.

interesting modification of the system is shown in Fig. 215 for the protection of a single phase transformer; this is an example of what may be termed current balance in the pilot circuits as against the pressure balance used in the foregoing example. The two instrument transformers are now arranged so that equal currents tend to flow in their secondaries for any particular load on the main power transformer, thus, when the two secondaries of the instrument transformers are connected so that at any instant the two currents tend to circulate in the same direction through the pilot wires, it is possible to select adjacent points on the two pilot wires, as a and a_1, between which there will be practically no difference of potential. If the relays which are to trip the switches are connected across these two points there will be no current

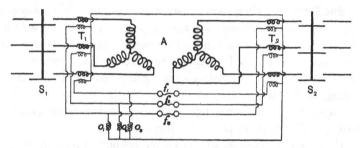

A. Main power transformer. T_1, T_2, etc. Current transformers for Merz Price Gear. S_1, S_2. Oil break switches. O_1, O_2, O_3. Relay coils for tripping the main switches. f_1, f_2, f_3. Fuses in pilot wires to give over load protection.

Fig. 216.

through them so long as the main transformer is sound; if, due to an earth or short in the main transformer, the currents in the secondaries of the two instrument transformers no longer tend to be equal, a difference of pressure develops between the points a and a_1 on the pilot wires, and current flows through the coil of the relay circuits resulting in the tripping of the main switches.

A similar system of current balance for a three phase transformer is shown in Fig. 216, in this case over load protection, in addition to protection against faults and earths, is secured by the addition of the fuses f_1, f_2 and f_3 in the pilot wires. As load increases on the transformer the current in the pilot wires also increases with the ultimate result that these fuses are blown, the current from one of the instrument transformers then of necessity circulates through the relay coils thus causing the tripping of the main switches. It is important to note that when pressure balancing is adopted, a break in the pilot wires will prevent the operation of the relays but is not immediately indicated and it is therefore necessary to frequently test the continuity of such wires; with current balancing a break in the pilot wires will cause the switches to trip, thus calling attention to the trouble. When used for transformer protection the two Merz Price transformers on any one line will not be exactly similar but will of course be arranged to take into account the fact that the normal currents on the high and low tension sides are not equal*.

The Ferranti Field system of feeder protection†. This is a system which has been developed for the purpose of cutting off a feeder in the event of the occurrence of an earth. It depends upon the fact that so long as a feeder, and the cables and apparatus connected thereto, are well insulated, the algebraical sum of the currents flowing in the three cores of the feeder will be zero and thus if a magnetic circuit is arranged to embrace the three cores no magnetic flux will be produced therein.

If an earth develops in the feeder beyond the point of protection, or in the system fed by the feeder, the algebraical sum of the currents in the three cores will no longer be zero (arrangements must be made so that the current leaking to earth does not return to the station by a path passing through the magnetic circuit of the device, thus, if the cable is provided with an earthed copper or lead sheathing, it is necessary to interrupt this when passing through the magnetic circuit and the necessary connection between the two portions made by a wire lying outside the magnetic circuit), a flux will be produced within the iron ring and an induced pressure, which may be used to operate the relay provided for opening the main switch, in a secondary wound round the ring. It will

* The merest outline of the Merz Price system is given above and for full information on the matter readers may consult a paper by Messrs Faye Hanson and Harling, *J. I. E. E.*, Vol. 46, p. 671.

† The writer is indebted to Messrs Ferranti for information concerning both the Ferranti Field and the Merz Price systems.

be clear that the device will operate for an earth occurring any-
where beyond the point where the magnetic circuit encircles the
cable and, unless care is taken in the application of the device,
sound cables may needlessly be cut off. In the case of parallel
feeders, for example, it would probably not discriminate between
the sound and the unsound cable in the event of an earth occurring
on one of them but would cause both to be switched off. It is
necessary (except perhaps in very large systems) that the neutral
point be earthed through a limiting resistance when this system
is applied.

The split conductor system for the isolation of faulty feeders.
This system depends upon the fact that if one conductor of a cable
is divided into two parallel cores with equal cross sections they

will have practically equal impedances (if due care is taken in the
manufacture and jointing of the
cores), and the current flowing
through the conductor will divide
itself equally between the two
paths. In order that the system
may be applied it is necessary to
divide each core of the cable to
be protected into two parts which,
in a cable for laying in the ground,
may be insulated as shown in Fig.
217. It is not necessary for the

Fig. 217. Three Core Split Conductor Cable [20,000 volts].

two parts of any one core to be insulated from each other to
an extent necessary to withstand the full pressure between the
lines of the system because, with normal conditions, there will be
no difference of pressure between the two parts, and even when
a fault develops the difference of pressure between the two parts
(due to the "drop" occurring in the paths caused by the out-of-
balance current brought about by the fault conditions) is not
likely to be nearly so large as the pressure between lines or between
a line and earth. Thus, in paper insulated cables for 20,000 volts,
the thickness of the paper between the two parts of one core may
be 0·1 inch, and for overhead lines, when suspension type insulators
are used, the two paths may be hung from the same string of

insulators, being separated only by a single link. The arrangements at either end of the cable to be protected are shown in Fig. 218, from which it will be seen that the two parts or "splits" of one core of the cable each pass round the primary windings of a special current transformer, the directions of the two primary windings being such that with equal currents in the two splits the resultant magnetic effect on the core of the transformer is zero. If an earth occurs on one split of the core the currents in the two sides will no longer be equal (in fact, as shown in the figure, the current in one side might possibly be reversed) and there will be a resultant magnetic flux threading the iron ring and this will produce a pressure in the single secondary winding which may be used to operate a relay and trip the switch controlling the feeder in question. It might be thought that if an earth developed on

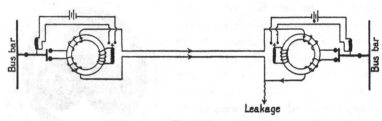

Fig. 218.

each of the "splits" simultaneously the relays would not operate the switch, but, in order that this should occur, it would be necessary for both earths to develop at the same instant and, further, at each instant during their continuance to pass equal currents (within a few amperes) to earth, and in practice it is almost inconceivable that this should happen. Experience has shown that the difference in the currents flowing in the two sides of the split when the cable is sound is very small, being frequently less than 0·1 % of the total current flowing, and it is thus permissible to set the relays to operate with but little out-of-balance current. It is clear that a short between either split of one core and either split of another core would also result in the operation of the relay. When the device is used on three phase systems of supply the neutral point of the system should be earthed through a limiting resistance and separate relays should be placed on each

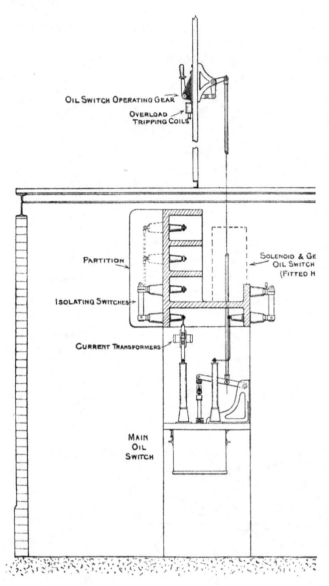

OIL SWITCH OPERATING GEAR

OVERLOAD
TRIPPING COILS

PARTITION

SOLENOID & GE
OIL SWITCH
(FITTED H

ISOLATING SWITCHES

CURRENT TRANSFORMERS

MAIN
OIL
SWITCH

Fig. 219.

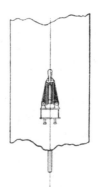

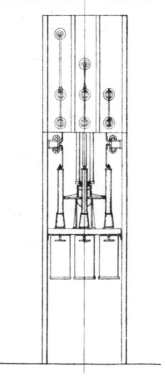

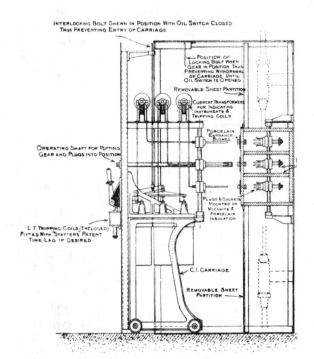

INTERLOCKING BOLT SHEWN IN POSITION WITH OIL SWITCH CLOSED THUS PREVENTING ENTRY OF CARRIAGE.

POSITION OF LOCKING BOLT WHEN GEAR IN POSITION THUS PREVENTING WITHDRAWAL OF CARRIAGE, UNTIL OIL SWITCH IS OPENED

REMOVABLE SHEET PARTITION

CURRENT TRANSFORMERS FOR INDICATING INSTRUMENTS & TRIPPING COILS

OPERATING SHAFT FOR PUTTING GEAR AND PLUGS INTO POSITION

PORCELAIN ENTRANCE BUSHES

PLUGS & SOCKETS MOUNTED ON MICANITE & PORCELAIN INSULATION

L T TRIPPING COILS (ENCLOSED) FITTED WITH STAFFERS PATENT TIME LAG IF DESIRED

C.I. CARRIAGE

REMOVABLE SHEET PARTITION

SIDE ELEVATION - WITH SWITCH CARRIAGE PARTLY WITHD.

E.H.T. SAFETY DRAW-OUT CUB

Fig.

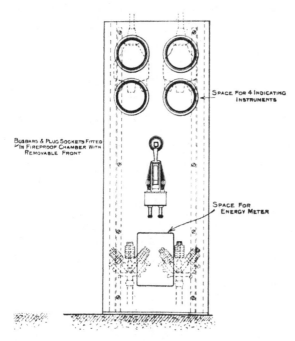

SPACE FOR 4 INDICATING
INSTRUMENTS

BUSBARS & PLUG SOCKETS FITTED
IN FIREPROOF CHAMBER WITH
REMOVABLE FRONT

SPACE FOR
ENERGY METER

RAWN   FRONT ELEVATION

ICLE (WORKING CAPACITY UP TO 2500 K.W.)

220.

phase in order that an earth on any one phase shall result in the tripping of the main switch. In the event of a break occurring in one of the splits, as may happen in the case of an overhead transmission line, the relays will operate and cut off the line and in all probability the falling wires would be made "dead" before they reached the ground. Another great advantage of the arrangement is the absence of pilot wires. The author is greatly indebted to The Electrical Improvements Co. of Newcastle for information concerning the system and for the diagrams shown in Figs. 217 and 218.

General construction and arrangement of high tension switchgear. The general appearance and arrangement of a high tension switchgear system depend much upon the type adopted, and the type again is largely influenced by the space available. In any case it is very necessary to prevent the formation of arcs between parts of the gear having different potentials and this is very commonly effected by adopting a cellular construction, separate compartments being provided for each bus bar and for each important part of the gear generally, the sides of the compartments being composed of non-inflammable material as concrete, brickwork, slate or metal. The insulating material most in evidence is porcelain at the points of support and air elsewhere, though occasionally bus bars and other connections are surrounded by a solid insulating compound; connections are best made of bare copper straps. Compactness, though in many cases of great importance, must never be attained at the expense of safety and reliability. Arrangements should be such that there is no risk of operators coming into contact with any live parts, these being well guarded when in use; it is also important that all parts which are normally alive should be capable of being made "dead," by isolating switches or other suitable means, for cleaning, inspection and repair.

The general arrangement of a switch board, operated by mechanical remote control and suitable for a high tension supply, as made by Messrs Switchgear and Cowans, to whom the author is indebted for the diagram, is shown in Fig. 219. The diagram is practically self explanatory but particular attention may be directed to the separation of the bus bars by the horizontal partitions and of the isolating switches by the vertical partitions.

A very interesting modification of a high tension switchgear arrangement, in this case suitable for direct control, is shown in

Fig. 221.

Fig. 220, which has also been supplied to the writer by Messrs Switchgear and Cowans, and which represents a board arranged on the draw out cubicle system.

Fig. 222.

The novel feature of the arrangement is that connections are made from the switch to the bus bars and lines by means of plugs and sockets suitably insulated, and it is thus possible to withdraw

21—2

the switch unit and its accessories completely from all live parts for cleaning and repair purposes. In this construction the plugs serve as isolating switches and a bolt is provided on the switch unit which interlocks with the frame and prevents

(1) the switch being withdrawn when closed, and

(2) the switch being replaced when closed.

A further example of the steel cubicle system is shown in Figs. 221 and 222, for which and for information concerning the same the writer is indebted to the General Electric Co.; here are shown photographs of standard panels as made by that Company for generator and feeder control. Perhaps the most distinguishing feature of the gear shown is that the bus bars and isolating switches are in a compartment, situated at the top of the panel, which is separated from the rest of the gear by sheet metal partitions. The isolating switches are operated by a hand wheel on the front of the panel and when these are open the main switch and all the rest of the gear except the isolating switches and the bus bars are "dead." Another very interesting and important part of the gear is the system of interlocks which is such that:

(1) The isolating switches cannot be opened when the main switch is closed (in addition to acting as a safeguard in the usual way this also prevents the isolating switches being blown open by electromagnetic action under short circuit conditions).

(2) The oil switch cannot be closed with the isolating switches open.

(3) The door at the back cannot be opened with the isolating switches closed.

(4) The isolating switches cannot be closed with the door at the back open.

Porcelain and air are the insulating materials used throughout and it will be realised that such gear as this requires no special foundations and can be fully connected up internally before leaving the works.

Lightning Arresters.

Transmission lines and cables may become charged to unduly high potentials owing to many causes, and these potentials may be at a very high frequency though occasionally they are direct rather than alternating in nature. These high pressures are likely to break down the insulation between the line and earth and, if they are allowed to penetrate to the windings of appliances connected to the line, a breakdown of the insulation may also be effected there. Amongst such disturbances as are indicated above may be included:

(1) A direct stroke of lightning on to an exposed part of the line. The effects produced in the immediate neighbourhood affected by the stroke are very severe and at present it does not seem possible to prevent damage to the line in the event of such a phenomenon occurring.

(2) Electrostatic induction resulting from a stroke of lightning. A charged cloud above a transmission line may induce a charge of opposite nature on the line and when the cloud is discharged, by a flash, the charge which has been induced on the line will become free and will raise the line to an excessive potential. The effects in this case are not, as a rule, so severe as are the results of a direct stroke and it is possible in most cases to adequately protect the line against trouble arising from this cause.

(3) Electromagnetic induction arising from a stroke of lightning. If a lightning discharge occurs in a path which lies parallel to overhead conductors it is possible that a considerable pressure may be induced in them due to the cause mentioned above. The three causes already mentioned are likely to give rise to disturbances of very high frequency.

(4) Electrical resonance. If the line has a considerable capacity it is possible that this, acting in conjunction with the inductance of the line or of the machines, may produce considerable pressure rises if the frequency of the line pressure is of a suitable value. Calculations and experience have shown that in the case of systems at present existing the fundamental frequency is not high enough to produce this effect, but of course there is always the risk of a high harmonic in the wave giving rise to the phenomenon. Pressure rises due to this cause will be of an alternating nature and will be of moderately high frequency.

(5) A variety of causes producing what have been termed surges, such as switching on cables through inductance, breaking a short circuit, earths or sudden alterations of load. The

consideration of such is not, at present, a matter for an elementary book and it is perhaps sufficient to mention their existence.

(6) Steady electrification due say to wind friction.

Appliances connected to lines are protected from trouble arising from the above causes by means of kicking (or choking) coils and lightning arresters. The choking coils used consist of a few turns wound inductively, but without an iron core, and which are connected in series with the line at the point at which it enters the station or sub-station. The magnitude of the inductance used, though sufficient for the purpose required, is too small to produce any appreciable impedance to the line current at normal frequency. The impedance offered to the disturbances is due to the high frequency and high rate of change of current with which they are commonly characterised.

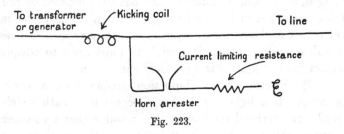

Fig. 223.

The arrester is connected on to the line side of the choking coil and is arranged to give an easy path to earth for the abnormal charge on the line and, at the same time, prevent a discharge to earth occurring due to the ordinary line pressure. After an arrester has discharged to earth, due to an abnormal pressure occurring on the line, there will always be a special tendency for the line pressure to maintain the discharge and perhaps one of the most important functions of the arrester is to quench this as rapidly as possible.

The horn arrester. This is one of the simplest types of arresters and consists of a spark gap, in series with a suitable current limiting resistance, connected between each line and earth. The length of the gap is chosen so that the normal pressure between the line and earth is insufficient to produce a discharge. When a discharge has occurred, owing to an abnormal pressure, the line pressure is amply large enough to maintain the arc unless proper arrangements are made for its extinction: in this case the arc rises to the top of the horns, partly due to electromagnetic action and partly due to a thermal effect, and eventually breaks.

Use of condensers. Messrs Isenthal and Co. (to whom the author is indebted for the diagram shown in Fig. 224 and for information concerning the same) have brought out a type of arrester which has proved to be very efficient in dealing with high frequency disturbances. It consists of condensers of the Moscicki type (which are of course extremely suitable for dealing with high pressures) having one pole connected to earth and the other to the line to be protected. It is usual to provide a fuse in series with each condenser and a water switch in order that the condensers can be gradually switched on and off. It is interesting to note that while the capacity of each condenser is quite small it is often found that the fuses are blown during discharge without the condensers being damaged, a tribute to the high frequency of the disturbances dealt with.

The electrolytic arrester. Certain metals, when used as electrodes in appropriate electrolytes, have an apparently non-conducting film formed upon their surfaces when a suitable difference of pressure is applied; it has been found, however, that the resistance which this film offers to the passage of a current almost completely breaks down if the pressure between the two electrodes is sufficiently increased. A metal which exhibits this phenomenon to a considerable extent is aluminium and when in conjunction with a suitable electrolyte the film formed will stand a pressure of the order of 450 volts. It will be evident that we may, by putting a sufficient number of such cells in series, construct a device that will prevent the ordinary line pressure passing a current through it and at the same time readily break down when an excessive pressure is attained between line and earth.

Other well known forms of arresters include the Wurtz pattern, which depends for its action upon the arc suppressing properties of bulky masses of certain metals, and the water jet arrester which operates by introducing a permanent leak on each line.

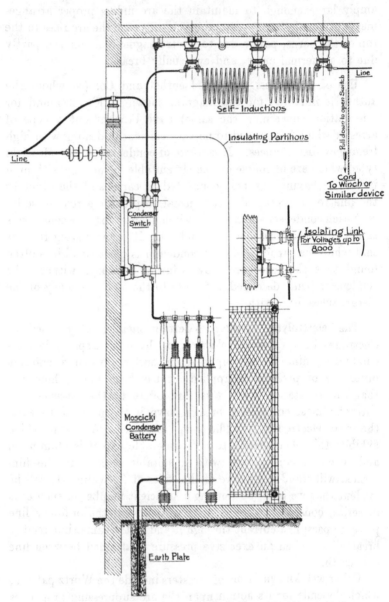

Installation of Moscicki Condensers for protection of 18,000 volt, 3 phase, overhead line.

Fig. 224.

In conclusion it should be stated that overhead lines may also be protected against the effects of lightning by means of a substantial galvanised iron wire, carried along the tops of the poles supporting the overhead wires, and earthed at frequent intervals, preferably at each pole*.

Many types of arresters have been and are used, and the question as to which is the best pattern is a very controversial one. This is perhaps the more so since an arrester has to deal with a variety of forms of disturbance and it is perhaps unfair to expect any one type to deal satisfactorily with all conditions likely to be met with in practice. Thus a type which is very suitable for dealing with high frequency disturbances (as, for instance, the condenser type) is not also the best for dealing with truly static phenomena, and ultimately it may become common practice to instal different types of discharge devices on the same line to meet the different conditions which may arise.

Material for the Conductors of Transmission Lines.

In general the choice of the material of the conductor for lines lies between:

(1) Hard drawn copper,
(2) Annealed copper,
(3) Aluminium.

For overhead lines the choice lies between hard drawn copper (used in preference to annealed copper on account of its greater tensile strength) and aluminium†.

The resistance of aluminium is, for a certain section and length, about 1·64 times that of copper, but for equal temperature rises the section of aluminium required to carry a certain current is only 1·45 times that of copper, the discrepancy between the above figures being accounted for by the greater radiating surface of the aluminium conductor. The relative weights of aluminium and copper for equal conductivities are as 1 is to 2, and, since the relative costs per pound are as 10 to 7·5 (approximately), it follows, as far

---

* For full information concerning the protection of lines and appliances against the effects of excess pressure the reader may consult papers by Mr J. S. Peck, *J. I. E. E.*, Vol. 40, p. 498 and M. Leblanc, *J. I. E. E.*, Vol. 51, p. 701.

† See paper by Mr Pannell, *J. I. E. E.*, Vol. 49, p. 817.

as the conductor is concerned, that a saving in first cost can be effected by adopting aluminium. A further advantage in favour of aluminium for very high tension lines is that the corona loss is not so troublesome on account of the greater diameter. The disadvantages of aluminium lie in its greater coefficient of expansion with temperature, greater area of surface exposed to the wind, lower tensile strength, difficulty in jointing, and the possibility of corrosion, some of which necessitate the use of shorter spans than would give the required factor of safety in the case of copper. To guard against the corrosion of aluminium it is desirable to arrange that no metal other than aluminium comes in contact with the wires, and the difficulty of jointing is got over by the adoption of mechanical joints of the clamped or torsion patterns. Stranded conductors are likely to be employed either with copper or aluminium, and, in conclusion, it may be said that there is a considerable prospect of the extensive use of aluminium for overhead lines even when the extra cost of the poles and insulators is taken into account.

For underground continuously insulated cables there are perhaps not the same advantages to be gained by the use of aluminium, because the saving in the cost of the conductor is likely to be largely or completely neutralised by the extra cost of the insulation, lead covering and armouring, due to the larger diameter of the conductor, and thus annealed copper is most commonly employed.

Comparison of overhead and underground transmission. For the transmission of electrical power there is the choice of using bare conductors as overhead lines, porcelain being used for insulation at the points of support, and continuously insulated cables laid underground. The former system will have the great advantage of lower first cost (on which undue stress is perhaps sometimes laid to the exclusion of proper consideration of running and total costs) and this may open up possibilities of supply over longer distances than would otherwise be feasible. Again, overhead lines have a considerably smaller electrostatic capacity than equivalent cables which are laid underground, due to the greater distance between the wires and also to the less dielectric constant of the insulating material; this lower capacity is advantageous in diminishing charging current and also in diminishing the possibility of resonance troubles. Against these advantages must be set off the greater liability of overhead lines to trouble from external causes leading to greater inspection and repair expenses. In England much attention has been paid to the use of underground cables and they are in extensive use up to pressures of 20,000 volts; for higher pressures, at the present time, overhead lines would most likely be used. Abroad, where very high pressures have been used to a considerable extent and where legal restrictions are perhaps not so severe, the overhead line has made much progress and is very common.

Choice of section of conductor. In order to arrive at the most suitable cross-section of conductor for use in any case the problem must be examined from several points of view. Generally the best section to use will be the one involving the least total annual expenditure, that is the section for which the sum of the annual fixed costs (*i.e.* the annual cost incurred on account of interest on the original outlay, depreciation, etc.) and the annual running costs (*i.e.* the annual cost of the energy wasted in the cable) is a minimum. The section worked out from the above considerations may not, however, be permissible from other points of view of a more technical nature.

In the first place the section must be sufficiently large to prevent overheating; this of course is largely a question of current density, the maximum value allowable being dependent to some extent upon the magnitude of the current to be carried and to a very considerable extent upon the nature of the surrounding dielectric, being much greater, for instance, in the case of overhead lines than for paper insulated cables.

In most cases the minimum section dictated by other considerations will be ample from the current density point of view, the only exceptions likely to arise being in the case of very short transmissions and possibly in very high tension transmissions.

Another important technical consideration that settles a minimum permissible limit to the section of the conductor is the pressure drop in the cable. In the case of very long lines this is perhaps likely to be the most important consideration, involving the use of a section that is considerably greater than that necessary from either of the points of view already discussed. In calculating the pressure drop in a line it is necessary to take into account not only the resistance (due allowance being made for the skin effect should it exist to an appreciable extent) but also the inductance of the line. It is necessary also to take into account not only the load current on the line but also the current due to the capacity of the line, since this will not only alter the magnitude of the total current on the line but also alter the phase angle between line current and line pressure which will also have its influence on the resultant pressure drop in the line. Finally, the question of the loss of power and the efficiency of the transmission should be considered, and in this connection it is perhaps important to note that the power lost in the line is (apart from the effect of brush discharge, leakage, etc.) always given by the formula $P = I^2 R$.

High tension cables for underground use. The choice of the insulating material to be used lies between vulcanised rubber, vulcanised bitumen and impregnated paper, and at the present time the latter is by far the most common material met with, especially for pressures above 3000 volts. The advantages in favour of the paper insulation are cheapness, low dielectric constant (leading to low capacity), low dielectric losses and ability to withstand higher temperatures (leading to greater permissible current densities from the heating point of view).

Paper insulated cables. It is unnecessary to tin the strands of the core intended for a paper insulated cable, since the paper and its impregnating material do not tend to corrode the copper, and the paper used is wrapped on in spiral layers to a thickness depending upon the pressure for which the core is to be insulated. The best paper for the purpose is that made of pure Manilla fibre, free from admixture with straw and wood fibres, the thickness of the strip used for high tension cables being usually of the order of three or four mils, and the breadth perhaps a half or a quarter of an inch. The tensile strength is of the order of 9000 lbs. per square inch. The objects of impregnating the paper—which is best done before it is applied to the core—are to secure flexibility of the cable (by allowing layers of paper to slip over one another), preserve the paper and increase the dielectric strength. The compound used for the impregnation of the paper should be plastic at ordinary temperatures so that there will be the minimum tendency for it to drain away from any point, and this is of particular importance in the case of cables used in mining work which may be required to be placed in a vertical position down a shaft. It is interesting to note that the impregnation lowers the ohmic resistance and therefore high insulation resistance should not, in paper insulated cables, be regarded as a criterion of quality since it may simply indicate insufficiency of impregnating compound.

It is usual to build the three cores of a three phase underground line into one cable and the three conductors may be arranged concentrically with respect to each other (layers of the insulating paper being placed between each core), or they may be arranged so that their centres form the corners of an equilateral triangle when the cable is viewed in cross-section. The latter method is the more common and when this arrangement is adopted it is customary not only to place a layer of paper round each core but also to place a layer round the cores as a whole thus further insulating them from earth (see Fig. 225). Since paper, even when impregnated, will not resist moisture, it is necessary to use efficient means to prevent trouble due to this cause and this is usually effected by surrounding the cable with a continuous sheathing of lead placed on, as a rule, in two layers, the absence of flaws in the lead sheathing being tested by a combination of hydraulic pressure and insulation tests immediately after the sheathing has been put on. On top of the lead sheathing a wrapping of impregnated jute or Hessian tape is placed.

Many paper insulated cables are armoured in order to protect them from mechanical damage and this is provided for by a single or double layer of steel wires or tapes, the armouring being protected from corrosion by further wrappings of impregnated jute or other suitable material. If the cables are for direct laying in the ground the armouring is essential and it adds materially to the mechanical strength of the cable even if it is intended to be laid in other ways. Since it is important to arrange that the ground, pipes in the vicinity of the cable, and any other conducting material in the vicinity of the cable shall not become electrified, it is customary to arrange that a metallic covering of the cable shall be made electrically continuous throughout (especial care is necessary at junction boxes, etc.) and efficiently earthed. The metallic covering occasionally consists of a copper lapping inside the lead sheathing, or the armouring may be used for this purpose or even the lead sheathing itself. In very high tension cables, having a considerable thickness of paper insulation, moisture may possibly penetrate through a defect which has developed in the lead sheathing and yet months may elapse before its presence is felt. In order to detect the presence of the moisture as early as possible, the British Insulated and Helsby Cable Co. have adopted the very interesting plan of placing just inside the lead sheathing, but separated from it by a thin layer of paper, an open spiral of copper or aluminium. Obviously the paper between the lead sheathing and the open spiral will be affected immediately moisture penetrates into the cable and the penetration is shown by a deflection being obtained on a galvanometer connected between the lead sheathing and the spiral due to a couple being set up.

Cross-sections of cables for various pressures are shown in Fig. 225.

Methods of laying underground cables. When cables are to be laid underground three well known methods are available:

(1) Armoured cables laid direct in the ground,

(2) Cables drawn into pipes,

(3) Cables laid on the solid system.

Of recent years the first method has tended to displace the others on account of its cheapness and flexibility, and the chief trouble to be feared from its adoption is corrosion of the metal coverings in unsuitable soils and damage caused by subsequent excavation. The risk of trouble on account of the latter cause may be minimised by placing, some 6 or 10 inches above the cable, a layer of tiles or stout boards which will give some small amount of mechanical protection and at the same time serve as a warning to excavators. In populous districts it may be preferred to adopt the more expensive system of laying the cables in pipes, which may be of cast iron, stoneware or fibre, but in any case the joints between successive lengths should be well made and every precaution taken to prevent moisture coming in more or less permanent contact with the cables. The third method consists of laying in

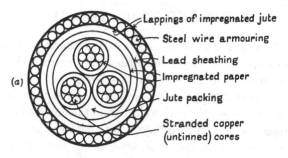

Armoured, lead covered, paper insulated cable for pressures
up to 3000 volts (Glover and Co.).

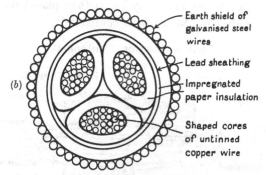

Lead covered, paper insulated cable for pressures up
to 11,000 volts (Glover and Co.).

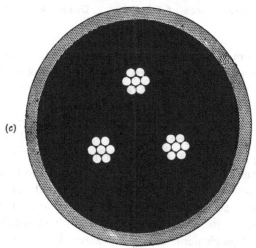

Cable for line pressure of 20,000 volts (British
Insulated and Helsby Cable Co.).

Fig. 225

a trench, which has been excavated to suitable dimensions, a trough composed of wood (which has been treated with a preservative material) or cast iron; in this trough, at intervals of say 2 feet, are placed supporting pieces of wood, asphalt or porcelain which serve to keep the cables clear from the bottom of the trough. The trough is then filled in layers with bitumen or some cheaper compound poured in hot, each layer being allowed to set before the next is poured on, the process being continued until the trough is practically full and the cable completely surrounded; a cover of iron or wood, as the case may be, is then placed on the trough which is finally topped up with compound*.

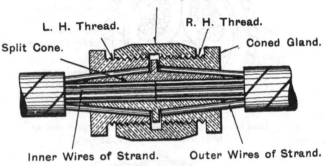

Fig. 226.

Joints in cables. The weakest points of an underground cable system are the joints between successive lengths of the conductor, and it is necessary to take the utmost care that not only shall the conductivity of the joints be equal to the conductivity of the cable, but that the subsequent insulation of the joint be well carried out. To get due electrical conductivity in the joint either a sweated joint or one involving the use of a mechanical coupling can be employed, in the former case it is very important to take every precaution that there shall be no subsequent corrosion of the joint due to any of the materials employed in its construction. A type of mechanical grip for a straight through joint, as made by Messrs Glover and Co., is shown in Fig. 226 which is self explanatory.

* See paper by Mr Vernier, *J. I. E. E.*, Vol. 47, p. 313.

The joint having been properly made and insulated with a material appropriate to the type of cable used, it becomes necessary to surround it with an arrangement which will effectively prevent the subsequent entry of moisture. This is commonly effected by surrounding the joint with a case which is afterwards filled with joint box compound, poured in hot and then allowed to solidify. Since most joint box compounds contract a little on solidifying it is necessary to take every precaution to ensure the final complete filling of the box.

The containing case may take the form of a lead sleeve which can readily be made watertight, or a cast iron box may be used as

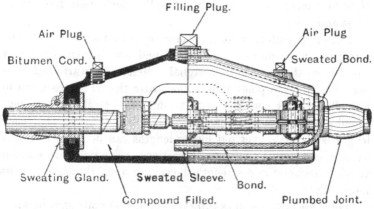

Fig. 227.   Junction Box.   (Messrs W. T. Glover and Co.)

indicated in Fig. 227, which shows a connecting box for connecting two lead covered concentric cables. It will be noticed that a bond is placed across the box in order to ensure electrical continuity of the sheathing. The writer is greatly indebted to Messrs W. T. Glover and Co. and to The British Insulated and Helsby Cable Co. for information and diagrams concerning the construction of cables for use underground.

Overhead line construction. The main support for the overhead line may consist of:

  (1)   Wood poles, either of the single or compound types,
  (2)   Steel poles,
  (3)   Reinforced concrete poles, or
  (4)   Steel lattice work towers.

The first three are suitable for small and medium sized transmission lines having spans up to eighty or one hundred yards, and in this country are said to be cheaper than built up towers under these conditions. In England the timber chiefly used for poles is red fir and in order to ensure a long life (say 20 to 30 years) it is necessary that it be properly seasoned and treated with creosote or other preservative material; especial care should also be taken to prevent the penetration of moisture at the top of the pole. On account of the varying quality of individual poles, and also on account of the risk of deterioration, the Board of Trade insist on the use of a large factor of mechanical safety, this being taken as 10 with a lateral wind pressure of 30 lbs. per square foot on the wires and poles.

Latterly, reinforced concrete poles have attracted considerable attention on account of the long life and low upkeep cost anticipated from the use of this material. Messrs Siegwart, Ltd., who manufacture such poles, inform the writer that they are made by helically winding concrete, with an immersed steel reinforcement, round a hollow steel core thus forming a hollow tube. The usual length is about 31 feet but longer lengths can be made if desired by joining together two or more sections. The poles are made of varying diameter and strength depending upon the class of work for which they are intended.

For high pressure lines (upwards of 40,000 volts say) in which considerable separation of the lines is a necessity, and when supports of considerable strength are required, steel lattice work towers are most commonly employed, spans of the order of 200 yards being used. In America, where such lines are numerous, they have become the standard method of line support. The writer is indebted to Messrs Buller and Co., Ltd. for information concerning steel towers and also for the material used in the production of Fig. 228 which represents, in skeleton, an intermediate tower as used by the Victoria Falls and Transvaal Supply Co. in South Africa.

Practically the whole structure is composed of angle iron made from Siemens Martin steel, the various members being held together by bolts. To prevent corrosion all the parts are, before erection, brushed with steel brushes (to get rid of scale and rust) and then galvanised or treated with bitumastic solution. On account of the considerable bulk of the towers it is necessary that each be put together at the point of erection and, to facilitate the subsequent raising of the tower, it is usual to arrange a pivot joint immediately above the ground; the tower can then be put

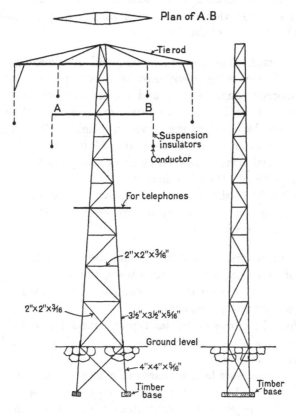

Fig. 228.

together in a horizontal position, raised to the vertical position, and duly secured. It will be noticed from the diagram that the width of the tower in the direction of the length of the line is less than in a direction at right angles to the length of the line, this is due to the fact that so long as the line remains sound the chief stresses to be feared at an intermediate tower are those due to cross winds. Towers used at terminations of lines, and at angles, are of a stronger type and are provided with a square base. In some cases the small breadth of base in the direction of the line has been carried to extremes and what has been termed a flexible construction (in the direction of the line) adopted*. Such flexible poles have the advantage of materially reducing the stresses on the poles in the vicinity of broken wires owing to their yield along the direction of the line.

Cross arms. These may be composed of creosoted wood or steel, the former being chiefly used for wooden poles and the latter for reinforced concrete poles and for steel lattice work towers. The height at which the cross arms are fixed is settled in this country by the fact that the minimum height of the wires above the ground is ordinarily 22 feet and 25 feet at road crossings (where special arrangements have also to be made to prevent risk of damage should the wires break).

The height of the cross arm will obviously need to be greater when suspension type insulators are employed than when pin type insulators are in use.

Line insulators. The insulators for use on overhead lines are made of porcelain, glass or stone ware, the former being almost invariably used since it is cheap, comparatively strong and is not damaged by atmospheric action.

Before describing particular types of insulators it will be well to consider a few general requirements which need to be met, these are as follows†:

(1) There must be a sufficient thickness of porcelain between the line wire and the earth connected supports of the insulator to prevent the line pressure causing failure of the insulator by piercing the porcelain. Since it is not possible to make a thickness of porcelain greater than a half or three quarters of an inch which can be depended on to be homogeneous throughout, it is customary to use two or three thicknesses of porcelain between the line and

* See paper by Messrs Matthews and Wilkinson, *J. I. E. E.*, Vol. 46, p. 562.
† A specification for high tension porcelain insulators is given by Messrs Matthews and Wilkinson, *J. I. E. E.*, Vol. 46, p. 580.

the earth connected support when high pressure is in use, these pieces being fired separately.

(2) The porcelain used should be thoroughly vitrified throughout, be practically non-hygroscopic (porcelain which is used for low tension work, as switch bases, is frequently very hygroscopic) and should be glazed (to keep the surface of the insulator clean) over the whole of the surface except such parts as are afterwards to be covered with cement.

(3) The general design of the insulator should be such that the risk of surface leakage is a minimum. To keep the surface leakage down the leakage paths should be long and have small breadth; they should also be kept as dry as possible under all weather conditions. The above considerations point to the use, in pin type insulators, of narrow and deep inner petticoats with a wide outer petticoat to keep the inner ones dry. On account of the high specific resistance of porcelain, leakage through the body is not likely to cause trouble.

(4) There should be an ample length of path, through the air, from line wire to pin to ensure the prevention of glow and brush discharge (which may involve considerable loss of power and ultimately lead to the production of an arc). In pin type insulators this consideration points to the use of a wide petticoat, and in suspension type insulators the difficulty is met by using a sufficient number of insulators in series. It should be noted that conditions (3) and (4) can both be met in pin type insulators by the adoption of a wide top petticoat with narrower ones underneath.

(5) The insulator and its supports must be sufficiently strong mechanically to withstand the lateral forces due to wind pressure, the vertical forces due to the weight of the wire, and, at terminal positions and bends, the longitudinal forces due to the pull of the wire.

In conclusion it should be stated that the mechanical arrangements of the insulator should be such that it is not likely to be pulled down in the event of a line wire breaking, and this may be effected either by making the insulator and its supports of sufficient mechanical strength to withstand the forces called into play under those conditions or by making arrangements so that the line wire

may slip relatively to the insulator should the former break, thu
preventing the insulator from being pulled down. In some case
the wires have been tightly fastened to the towers at intervals
where stronger towers are placed, and but lightly secured at th

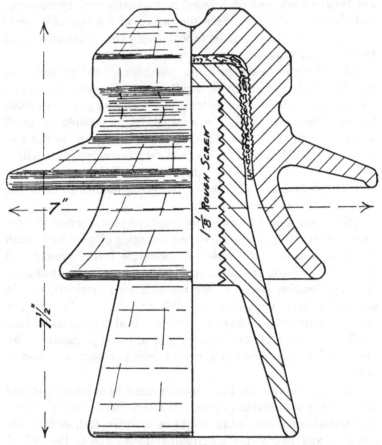

Working pressure 10,000 to 15,000 volts. Rain test 55,000 volts.
Dry test 75,000 volts. Side pull 1000 lbs.

Fig. 229.

intermediate and weaker towers and supports. Further points in
connection with the securing of the maximum mechanical strengtl
in insulators consist in utilising the porcelain in compressiv
stress only, and in having the insulator of neutral tint in order tha

it may not serve as such an attractive target for stones and other projectiles.

All the diagrams of insulators shown on pp. 342—348 have been kindly supplied by Messrs Buller and Co., Ltd., of Tipton, Staffs., to whom the writer is also indebted for information concerning the capabilities of the same.

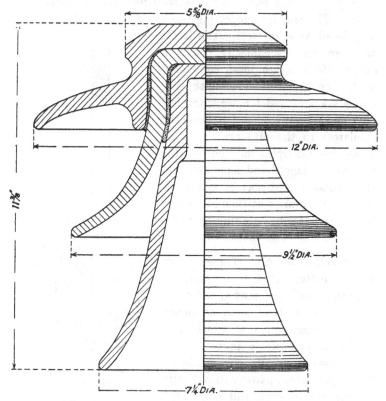

Working pressure 40,000 to 50,000 volts.   Rain test 100,000 volts.
Dry test 135,000 volts.   Side pull 2000 lbs.

Fig. 230.

Pin type insulators.   These may be used on transmission towers and poles up to pressures of 60,000 volts, beyond which their increased size and cost renders their use uneconomical. Examples of such insulators are shown in Figs. 229 and 230 and in addition to the information as to capabilities given under each diagram the following general points are of interest.

(1) The transmission wire may pass through the top groove or through the side groove, being attached thereto by a suitable binding.

(2) The insulator is made in two or three parts, for reasons already given, the parts being fastened together by cementing.

(3) The pins for supporting the insulator from the cross arm may be of wood or steel. The latter are stronger mechanically, have a longer life and are most commonly used. The insulators shown in the diagrams are arranged for use with steel pins, and several modes of securing the insulators to the pins are available. Thus, for example, the end of the pin may be jagged and cemented into the insulator, or it may have a screw thread cut on and may then be secured to the insulator by screwing in with a hemp packing (Fig. 231).

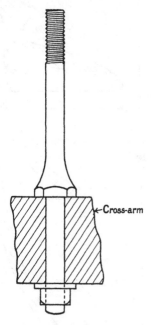

←Cross-arm

Insulator pin for screwing in with hemp (Buller).

Fig. 231.

Shackle insulators. These are obviously better fitted to withstand a horizontal pull and may therefore be used at the ends of lines, at angles, and at other points at which large horizontal pulls are expected. They are, however, usually only made for use on pressures up to say 10,000 volts, on account of cost, and above this pressure the strain type of link insulator should be substituted.

Link type insulators. These are almost invariably used when the pressure exceeds 60,000 volts, the necessary sparking distance between the line wire and earth being obtained by connecting a sufficient number in series. Two main types may be distinguished:

(1) Those for use on intermediate towers and which are used as suspension insulators, the line joining the centres of the insulators in this case being vertical, and

(2) Those for use at the ends of lines, angles, and at strain towers and poles generally, and used in such a position that the line joining the centres of the insulators is approximately horizontal. Diagrams of link type insulators are shown in Figs. 233, 234, and 235 and an interesting feature of the construction in the last two cases is, that in the event of the porcelain of the insulator failing, the line wire is still held up owing to the interlocking of the upper and lower suspending wires of each insulator.

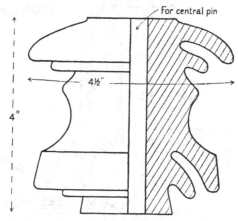

Shackle Insulator.   Working pressure 1000 to 2000 volts.
Rain test 8000 volts.   Dry test 25,000 volts.   Pull 3000 lbs.

Fig. 232.

Of course the insulating value of the link is, under these circumstances, reduced to zero, and a certain extra electrical stress is thrown on to the others in the same series.

In regard to the pressures which are given for each insulator a word of explanation is perhaps necessary as to the rain test; this is in each case taken with a shower of water falling at an angle of 45 degrees and of an intensity equivalent to a fall of one inch of rain in five minutes*.

* A very complete paper on porcelain insulators has been published by J. Lustgarten, *J. I. E. E.*, Vol. 49, p. 235.

The pull given for each insulator is that which causes a permanent set.

Arrangement, spacing, etc. of overhead wires. Three phase overhead lines are commonly arranged so that, when viewed end on, the three wires form an equilateral triangle, but occasionally the three wires are arranged in a vertical line (as, for instance,

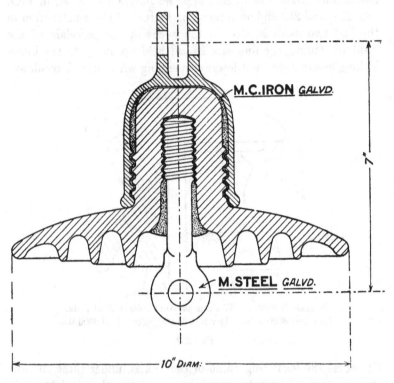

M.C.IRON GALVD.

7"

M.STEEL GALVD.

10" DIAM:

Link Suspension Insulator. Working pressure (single unit) 20,000 volts.
Dry test (single unit) 70,000 volts. Rain test (single unit) 52,000 volts. Pull 3 tons.

Fig. 233.

when duplicate lines are run on the same set of poles or towers) and also more rarely in a horizontal line.

The spacing between the wires should be sufficient to prevent trouble due to brush discharge, and also to prevent the wires coming too close together when swaying due to wind. Messrs

Matthews and Wilkinson state that a distance of one foot between
the wires for every 1000 volts seems to be usual.

The mechanical tension in an overhead line is proportional, for

LINK   STRAIN   INSULATOR

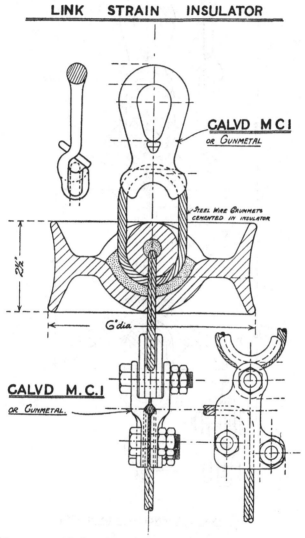

Working pressure (single unit) 10,000 to 15,000 volts.   Dry test 60,000 volts.
Rain test 30,000 volts.   Pull 1·5 tons.

Fig. 234.

a given wire, to the square of the length of span and inversely proportional to the dip of the wire between the poles, the exact

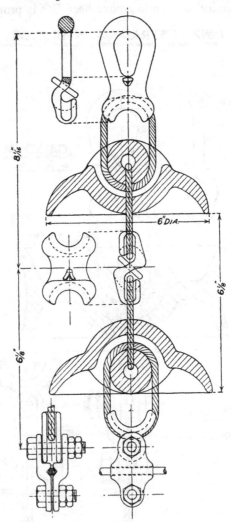

LINK SUSPENSION INSULATOR

Working pressure (single unit) 10,000 to 15,000 volts. Dry test 60,000 volts.
Rain test 30,000 volts. Pull 1·5 tons.

Fig. 235.

relationship being given by the formula $T = \dfrac{wl^2}{8d}$, where d is the
dip in feet, l the length of the span in feet, w the force on the
wire in a direction at right angles to the length of the wire in
pounds (this is partly due to the weight of the wire and partly
due to wind pressure) and T is the tension of the wire in pounds.
It should be noted that any small alteration in the length of the
wire in any one span exerts a great influence on the dip, and, further,

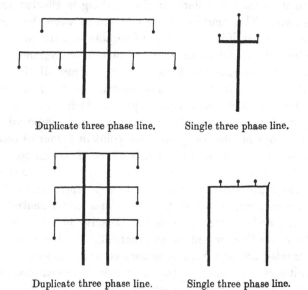

Duplicate three phase line.　　Single three phase line.

Duplicate three phase line.　　Single three phase line.

Fig. 236.

that small changes in the length of the wire in a span are continually being produced by changes in temperature and by changes
in the force acting per foot run of the wire caused by wind and
snow. The Board of Trade regulation in regard to the mechanical
safety in overhead wires states that there must be a factor of
safety of 5, assuming a lateral wind pressure of 30 lbs. per sq. foot,
and a temperature of 22° F., and, since the wires will not be erected
at such a low temperature, it is necessary to calculate the diminished tension suitable to the actual stringing temperature and
observe that this is not exceeded during the erection.

Earthing of three phase systems. Recently considerable attention has been devoted to the question as to whether three phase systems should be earthed in any way.

In general, the point to be earthed in a three phase system is the neutral point, and if this is done the normal pressure between any line and earth is kept down to 58 % of the line pressure though momentarily, should one of the lines go down to earth, the pressure between the other lines and the earth may rise to the full line pressure particularly if the earthing is effected through a resistance. This limiting of the pressure between the lines and earth may permit, if it is considered desirable, of a certain economy in the insulation, and, in the case of low tension systems, may be helpful in minimising the danger to life. Again, if the neutral point is earthed, it facilitates the automatic cutting off of defective lines. On the other hand with a system which is normally completely insulated it is possible to continue to run temporarily if an earth develops on one line. Another point in favour of complete insulation is that in small systems an earth connected person accidentally touching one of the lines would not receive so severe a shock as if the system was earthed at any point; an earth, even of high resistance, on one of the other lines would neutralise this advantage, and in large systems the capacity between earth and the other two lines would cause practically the full line pressure to be exerted across a high resistance contact between earth and the third wire. It is also worthy of remark that on earthed systems it is necessary to protect each line against over load but on unearthed systems only two lines need to be so protected. The subject has been exhaustively dealt with by Mr Peck (*J. I. E. E.*, Vol. 50, p. 150) and in the discussion following, and the general opinion at the present time would seem to be to earth the neutrals of large supply systems through resistances, while smaller schemes can be left with free neutrals. In low tension three phase distributing systems it is usual to earth the neutral point.

INDEX

Addition of alternating currents 22
Alternating current
  chemical effect of 17
  heating effect of 20
  instantaneous value of 4
  magnetic effect of 18
  magnitude of 20
  R.M.S. value of 20
  skin effect with 57
  three phase 82
  two phase 80
Alternators 142
  armature reactance in 162
  armature reaction in 163
  armature resistance in 161
  armature windings for 152
  automatic regulators for 170
  compounded 168
  high speed 146
  inductor type 143
  moving armature type 142
  moving field type 144
  parallel operation of 172
  pressure developed by 160
  rotary synchroniser for 178, 180, 183
  salient pole type 147
  short circuiting of 150
  slot insulation for 154
  smooth core type 148
  stator of high speed type 149
  three plane winding for 155, 157
  two plane winding for 155
Ammeters 98
  dynamometer type 106
  hot wire type 100, 101
  induction type 111
  moving iron type 99
Apparent ohm 42
Armature winding of alternators 152
Automatic protection of alternators 309
Auto-transformer 214
  as booster 219

Berry transformer 205
Boosting transformer 219

Cable, concentric 52
  high tension 332
  joint box for 337
  mechanical connector for 336
Capacity 46
  of cable 47
  unit of 46
Cascade control of induction motors 247
Choking coil 74
Circuit breaker, electrical remote control type 305
  fixed handle type 300
  free handle type 302
  mechanical remote control type 305
Closed circuit winding 80
Condensance 48
Condenser 46
  foiled paper type 49
  for use with lamps 75
  Moscicki type 50
Condenser type arrester 327
Converting plant 258
Copper loss 67
Core balance system of protection 318
Cross arm for high tension transmission 340
Cubicle switchgear 324
Current transformer 117
Cycle 7

Delta connection 85

Earthing of three phase systems 350
Eddy current loss 19
Efficiency of converting plant 259
Efficiency of induction motors 251
Efficiency of transformer 208
Electrical resonance 53
Electrolytic type arrester 327
Electrostatic type voltmeter 102
Energy measurement in multiphase circuits 90
Everett, Edgcumbe synchroniser 183

Farad 46
Feeder arrangement 312

Feeder switchgear 299
Ferranti Field type of protection 318
Flickering of lamps 8
Foiled paper type of condenser 49
Fourier's Theorem 12
Frequency 7
Frequency meter 118
    inductive circuit type 121
    resonance type 119

Generator, see Alternator
Generator switchgear 298

Harmonics 12
Henry 32
High tension cables 332
High tension switchgear 321
Horn arrester 326
Hunting of synchronous plant 267
Hysteresis loss 19

Impedance 44, 51
    of iron rail 60
Impedances in parallel 54
Inductance 29
    variation of with current 33
Induction motor 224
    air gap in 234
    as generator 253
    auto-transformer starter for 237
    cascade control of 247
    efficiency of 251
    form of slot in 232
    power factor of 250
    principle of three phase 225
    rotor of 232
    rotor resistance type starter for 240
    short circuiting and brush lifting
        gear for 242
    single phase type 255
    slip and torque in 252
    speed regulation of 244
    star delta type starter for 236
    starting of 235
    starting single phase type of 256
    stator of 229
    stator resistance type starter for
        239
    torque in 234
Inductive circuit 30
    flow of current in 40
    rise of current in 36
Inductor type alternator 143
Instrument transformer 115
Interconnection of three phase circuits
    82
Interconnection of two phase circuits 81
Isolating switches 309

Junction box 337

Kapp coefficient 161
Kicking coil 326
Kilo-volt-amperes 69

Lagging current 10
Lattice work towers 339
Leading current 10
Lenz's law 29
Lightning arresters 324
Lightning conductor 59
Line insulators 340
    link strain type 345
    link suspension type 345
    pin type 343
    shackle type 344
Linkages 30

Mechanical connector for cable 336
Mercury arc rectifier 295
    mode of action of 296
    starting of 296
Merz Price protection 315
Mesh connection 85
Microfarad 46
Moscicki condenser 50, 327
Motor converter 286
    pressure regulation of 291
    principle of 286
    speed of 289
    starting of 290
Moving armature type alternator 142
Moving field type alternator 144
Multiphase currents 79
    for lighting 88
    for motors 87
    for transmission 88
Mutual induction 188

Neutral point 85
Non-inductive circuit 30

Oil break switch 300
Ondograph 128
Oscillograph 131
    cathode ray type 135
    Duddell type 131
    hot wire type 135
Overhead line construction 337
    arrangement of wires for 349
    concrete poles for 338
    spacing of wires for 346
    towers for 339
    wood poles for 338
Overhead transmission 330
    section of conductor for 331
Over load protection 313

Paper insulated cables 332
Parsons' compounded alternator 168
Period 7

Periodicity 7
Phase 8
Phase difference 8
Phase transformation 220
Polyphase circuits 79
    power and energy in 90
Power, in A.C. circuits 62
    in multiphase circuits 90
    in non-inductive circuits 62
    in purely inductive circuits 63
    in three phase circuits 91
    in two phase circuits 90
    in unbalanced circuits 92
    wave form of 66
Power factor 68, 96
    table 70
Power factor indicator 122, 73
    for single phase circuits 123
    for three phase circuits 125
Pressure transformers 116
Puissancegraphe 130

Rating of alternators 69
Reactance 42
Regulation of alternators 166
Regulation of transformers 211
Relays for automatic switches 309
    Ferranti Field type 318
    Merz Price type 315
    over load type 314
    reverse energy type 311
    split conductor type 319
Representation of alternating currents 12
Resonance 53
Reverse power relays 311
R.M.S. value 20
Rotary converter 272
    pressure ratio in 275
    pressure regulation in 282
    principle of 274
    self synchronising type 281
    starting of 279
    wave form of current in 277

Self induction 29
    absolute unit of 32
    calculation of 32
    practical unit of 32
Self synchronising rotary 281
Short circuiting of alternator 150
Single phase current 79
    for lighting 88
    for motors 87
    for transmission 88
Single phase induction motor 255
Skin effect 57
Split conductor protection 319
Star connection 84
Starters for induction motors 235
Static balancer 293

Supply meters for A.C. circuits 135
    induction type 138
Synchronous motor 261
    effect of varying excitation of 265
    hunting of 267
    self starting type of 271
    starting of 270
    vector diagram for 263
Synchroscopes 180, 183

Three ammeter method 70
Three phase currents 82
Three voltmeter method 71
Tirrill automatic regulator 170
Transformers 188
    all day efficiency of 211
    auto-type 214
    cooling of 203
    copper losses in 207
    core construction in 195
    core losses in 196, 207
    core type 194, 202
    current ratio in 193
    efficiency of 207
    ideal 191
    magnetising current in 191
    phase 220
    pressure drop and regulation in 211
    pressure generated in 194
    pressure ratio of 190
    shell type 194, 204
    stampings for 197
    stepped cores for 198
    windings for 199
Trigonometrical formulae 13
Two phase currents 80
Two wattmeter method of power measurement 94

Vector diagrams 16
Volt-amperes 68
Voltmeter 98
    dynamometer type 106
    electrostatic type 103
    hot wire type 101
    induction type 111
    moving iron type 99
    Sumpner type 109

Wattless current 67
Wattmeter 72, 112
    dynamometer type 113
    phase error in 114
Wave diagram 15
Wave form 10
    determination of 126
    flat topped 10, 12
    peaked 12
    sinusoidal 10
Weston synchroscope 180
Windings for alternators 152